BANG!

THE COMPLETE HISTORY OF THE UNIVERSE

BANG!

THE COMPLETE HISTORY OF THE UNIVERSE

BRIAN MAY   PATRICK MOORE   CHRIS LINTOTT

CARLTON
BOOKS

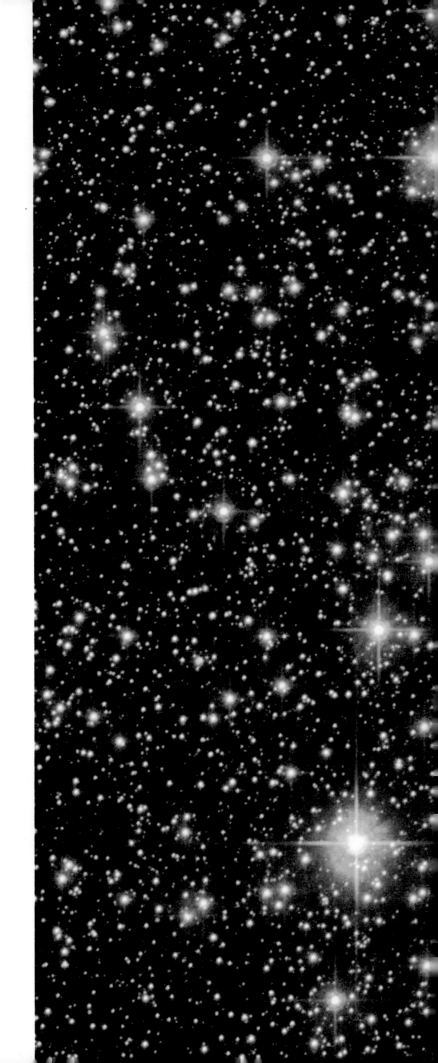

This third edition published in 2009 by Carlton Books Limited
20 Mortimer Street, London W1T 3JW
10 9 8 7 6 5 4 3 2

First published in 2006

Created by Canopus Publishing Limited, 27 Queen Square, Bristol BS1 4ND
www.canopusbooks.com

A CIP catalogue record for this book is available from the British Library.

ISBN: 978 1 84732 336 1
Printed in China

For Canopus:
Director Robin Rees
Major artworks Brian Smallwood
Diagrams James Symonds
Consultant editors Iain Nicolson, Garik Israelian
Project editor Julian Brigstocke
Text editor Tom Jackson
Art director Harriet Athay
Proofreaders Tom Short, Simon Laurenson

For Carlton:
Editorial director Piers Murray Hill
Art director Clare Baggaley
Publishing manager Penny Craig
Designer Emma Wicks
Production controller Lisa Moore

Image (this page): The lovely open cluster NGC 290 (C94), known as the
Jewel Box, lies in the same binocular field as the Small Magellanic Cloud.

▶ A Note on the Cover ◀

Our explosion on the cover is for fun only. There is no suggestion that
any part of the Big Bang ever looked like this. It would never be possible
for a human to stand outside the expanding Universe and take in such a
view, since outside the Universe, as we shall see, space and time do not
exist – there is no place to stand! However, we really have an even more
privileged position – we are all actually inside the mother of all explosions,
and the further we peer into space, the more we realize that, all around us,
everywhere we look, the Big Bang is what we see.

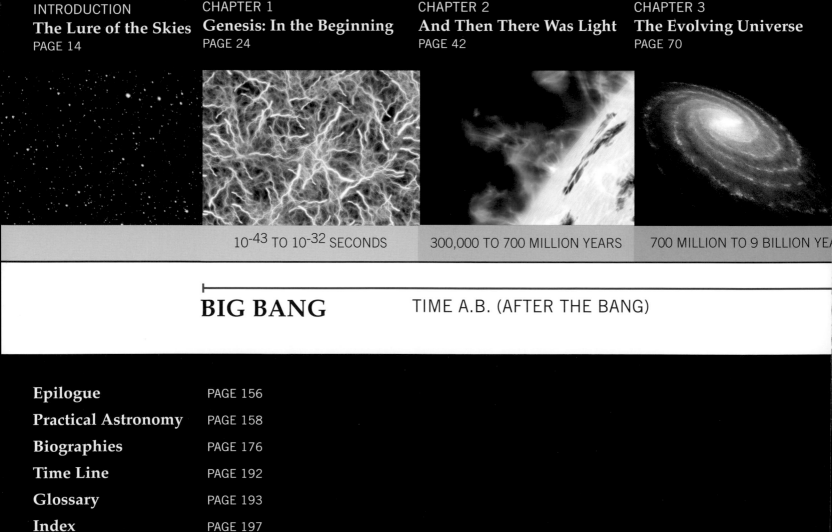

10^{-43} TO 10^{-32} SECONDS

300,000 TO 700 MILLION YEARS

700 MILLION TO 9 BILLION YEA

BIG BANG   TIME A.B. (AFTER THE BANG)

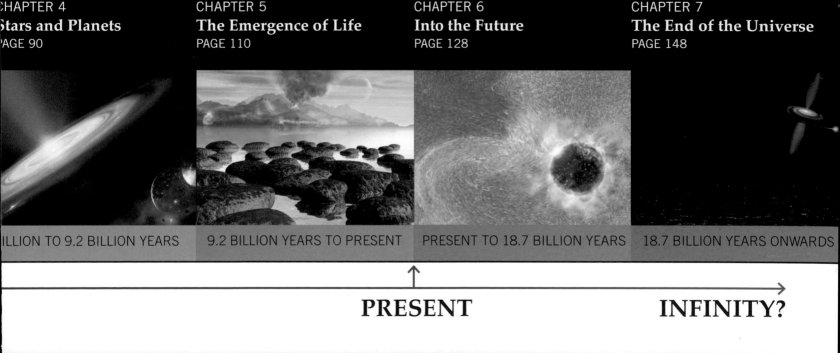

ILLION TO 9.2 BILLION YEARS

9.2 BILLION YEARS TO PRESENT

PRESENT TO 18.7 BILLION YEARS

18.7 BILLION YEARS ONWARDS

PRESENT

INFINITY?

PREFACE

None of us would be here discussing the so-called 'Big Bang', had it not been for the exclamation of an astronomer who considered the whole idea to be ridiculous.

From the late 1940s, eminent British astronomer Fred Hoyle was famously an advocate of what was called the steady-state hypothesis, proposed originally by Hermann Bondi and Thomas Gold. For philosophical reasons, Hoyle was attracted to this idea that the Universe on a broad scale ought to be unchanging with the passage of time. He and others noted that the component parts of the Universe were all fleeing away from each other, a discovery made by Edwin Hubble in the 1920s. So, to keep things looking the same, the steady-state theorists reasoned that new matter must be constantly coming into existence everywhere to replace the loss (a concept termed 'continuous creation'), with the result that the Universe would remain essentially the same forever. Meanwhile Ukrainian cosmologist George Gamow was arguing that, on the contrary, the Universe might have come into existence at a single instant, and was not in a steady state at all. In a radio broadcast in 1949, Hoyle strongly asserted that current observational evidence was in conflict with theories requiring all matter to have been created in 'one Big Bang', and in doing so unwittingly coined the name that henceforth would always be used to describe the theory he spent the rest of his life fighting against.

During the 50s and early 60s, a battle raged between the two theories, but gradually evidence began to pile up in favour of exactly the primeval explosion that Hoyle had found

◄ **The southern skies**

Seen from New South Wales, Australia, this circular 200-degree view of the night sky was taken with a fish-eye lens. Comet Hyakutake is at the top. The fuzzy patch towards the bottom is the Large Magellanic Cloud, a nearby dwarf galaxy in the process of being consumed by our own Milky Way, which is visible as a bright band. Looking at the brightest part to the left, we are looking towards the centre of our Galaxy.

◄ **The authors**

Chris Lintott and Brian May stand behind Patrick Moore as they prepare to observe the transit of Venus in 2004.

so unpalatable. Finally, in 1964, a mortal blow was delivered to the steady-state theory
when Penzias and Wilson (unwittingly at first) discovered the Cosmic Microwave
Background – the actual echo of the Big Bang itself, reverberating through the whole
of Creation, billions of years later.

The Big Bang theory, (or more precisely, collection of theories), is exactly what it says;
only a theory – a virtual model constructed to fit the available evidence that we find, as
we observe and measure the Universe we live in. In astronomy, models come and go.
The evidence is not yet all in, so we would be amazed if in a few years time our book
would not need to be substantially re-written. But the story we tell in these pages
constitutes what most astronomers currently believe to be a good model.

We have set ourselves the goal of telling our main story, the story of the evolution of
the Universe itself, in the order in which it happened, so we decided to put the historical
anecdotes and other diversions into what we have affectionately come to call 'grey areas',
rather than include them in the main text. If you wish to experience the story of the
Universe without interruption, feel free to skip the material in the grey areas; save them
for later. Bear in mind that our main story begins in Chapter 1. Each successive chapter
then describes the events of a period of time, up to the present day and beyond, into the
farthest reaches of the foreseeable yet almost unimaginably distant future.

At intervals in the top right-hand corner of the pages, you will find a helpful absolute
time reference – a reminder of how far we have moved along the time line. Our
convention in this book is to work on an absolute timescale with zero at the moment
of creation; for convenience, such times are notated as A.B. – After the Bang.

The discoveries we describe here were made by some remarkable pioneers in astronomy
and science. Their biographies are included at the end of the book, where there is also an
introduction to practical astronomy, written by Patrick; after all, we all started by gazing
up at the night sky, and wondering what it was all about.

For help and inspiration ...

Our thanks to Jimmy Alvarez, Tim Benham, Sara Bricusse, David Burder, Marcus
Chown, Adam Corrie, Jane Fletcher, John Fletcher, Garry Hunt, Roger Prout, Phil Webb,
Woodstock typewriters ... and of course Ptolemy and Jeannie.

Note on units

In keeping with current practice, we use the term 'billion' throughout to mean one thousand million.
Temperature is measured in degrees Celsius (Centigrade) or Kelvin (Celsius plus 273 degrees). To convert
Celsius to Fahrenheit, multiply by 1.8 and add 32.
The unit of astronomical distance is the light-year, equivalent to about 6 thousand billion miles or 9.6
thousand billion kilometres.

PREFACE TO THE THIRD EDITION

Since *Bang!* was first published in 2006, there have already been many exciting new developments in astronomy: for example, the surface of Mars has been explored in unprecedented detail, the search for dark matter in the Universe has intensified, Pluto has been reclassified, and many thousands of new images from ground-based and orbiting telescopes have amazed and intrigued. This new edition incorporates brand new material to reflect these recent discoveries, bringing our Complete History of the Universe Bang! up to date.

May, Moore, Lintott

Look upward on a dark, clear night and you will see stars – hundreds of them, even thousands, if you are lucky enough to live away from the light pollution of our modern cities. The heavens seem to be ablaze. Many people today know that these tiny, twinkling points of light are suns, many of them far larger, hotter and more powerful than our own Sun, and that our Earth appears to be an insignificant planet, perhaps less important in the wide Universe than a single grain of sand in the Sahara. But what lies behind it all? How did the Universe begin? How is it evolving, and how will it die, if indeed it ever comes to an end?

Astronomers are attempting to answer these questions, and it is truly remarkable that such minuscule beings, living on a small planet moving round an undistinguished star, have been able to peer into the depths of space, pick up light from star-systems unbelievably far away, and even send machines to other worlds. It may well be that other civilizations can outmatch ours, and that we must be regarded as cosmic primitives, but as a race we have at least made a start in coming to an understanding of the Universe we live in. In this book we are going to do our best to tell the story of the Universe – from its creation, long before the Earth existed, through to the present day, and then into the future, when the Earth will no longer be even a memory. There is much that we do not know, and perhaps will never know, but we have come a long way since our ancestors gazed up toward the stars, just as we do today, and wondered what they were.

We are living in a golden age of astronomy. New observational instruments such as the Hubble Space Telescope, which orbits the Earth beyond the haze of our atmosphere, would have been unimaginable only a few decades ago. Another factor which has been crucial in the amazing advances of the last fifty years is the increase in computer power available to scientists.

Nowhere has there been such spectacular recent progress as in the field of cosmology – the study of the past, present and future evolution of the Universe on the largest scale. During much of the 20th century, most astronomers favoured a static universe, uniform on the largest scales and largely unchanging over time. Our present picture could not be more different.

Where are we?

In our story we deal with immense distances and vast spans of time. The Earth, a globe about 8000 miles (12,800 km) in diameter, moves round the Sun at a distance of 93,000,000 miles (150,000,000 km); it is one of nine planets which, together with many more less substantial bodies, make up the Solar System.

Most of the planets have satellites; we have one, our familiar Moon, which is our faithful companion in space, and stays with us in our constant journeying round the Sun. Like the planets, it shines only by reflected sunlight; it lies a mere quarter of a million miles (400,000 km) from us, which is why it looks so imposing. It is the only other world that has been reached by human beings, and nobody who was alive in 1969 will ever forget the sensation of triumph as Neil Armstrong took 'one small step for [a] man, one giant leap for mankind' on to the bleak rocks of the lunar Sea of Tranquillity.

But the Solar System is a very minor unit in the Universe; our Galaxy, known as the Milky Way, contains at least a hundred billion (in this book one billion equals one thousand million) suns, and we know that many of these are attended by planets. We do not know whether these planets are home to any kind of life, let alone thinking beings.

▶▶ Trifid Nebula (M20)

The Trifid Nebula is a giant star-forming cloud of gas and dust. The bright filaments seen in this infrared image captured by the Spitzer Space Telescope are regions where stars are forming. They cannot be seen in observations sensitive only to visible light.

▼ Orion

The brightest stars of this magnificent constellation, always prominent in the night sky in winter, have suggested the strong shape of a man since prehistoric times; we know him as Orion the Hunter. The bright orange star, at the top left of the figure – his shoulder – is Betelgeux. The blue-white star Rigel, at bottom right marks a leg. Midway between them lie three stars almost in a line, known as Orion's Belt. The smaller line of stars running from the belt downward makes up Orion's sword, and includes the Orion Nebula, our nearest star-forming region, visible to the naked eye as a misty light around the central star of the sword.

At the speed of light

The stars are remote. To try to give their distances in miles or kilometres would be as clumsy as giving the distance between London and New York in inches, but fortunately there is a better unit to hand. Light does not travel instantaneously; it flashes along at the rate of 186,000 miles per second (300,000 kilometres per second), so that in one year it covers nearly 6 thousand billion miles (9.6 thousand billion kilometres). This is what has become known as the light-year (note that it is a unit of distance and not time). The nearest star beyond the Sun is just over 4 light-years away, while the most remote objects so far recorded are over 12 billion light-years from us.

Seen across such vast distances, the stars shine merely as tiny points of light. Appearances are deceptive; many of the stars visible on any clear night are not only much more luminous than the Sun, but also much larger. For example, Betelgeux in the constellation of Orion, over 300 light-years away, is vast. Its globe could contain the entire orbit of the Earth round the Sun. Some features have been detected on its surface, but only the Sun is near enough to be studied in real detail – and much of our knowledge of the stars in general depends upon what we have learned from studying our own neighbourhood star. Fortunately, the Sun is a very normal star, neither particularly powerful nor particularly feeble, and certainly not as variable as many. Astronomers rank it as a dwarf, but in fact it seems to be slightly more massive than the average – and giant stars such as Betelgeux are much less numerous than the dwarfs.

We can also gain a great deal of understanding by looking at the colours of the stars. Just as we talk of objects being red- or white-hot, with white-hot objects being hotter

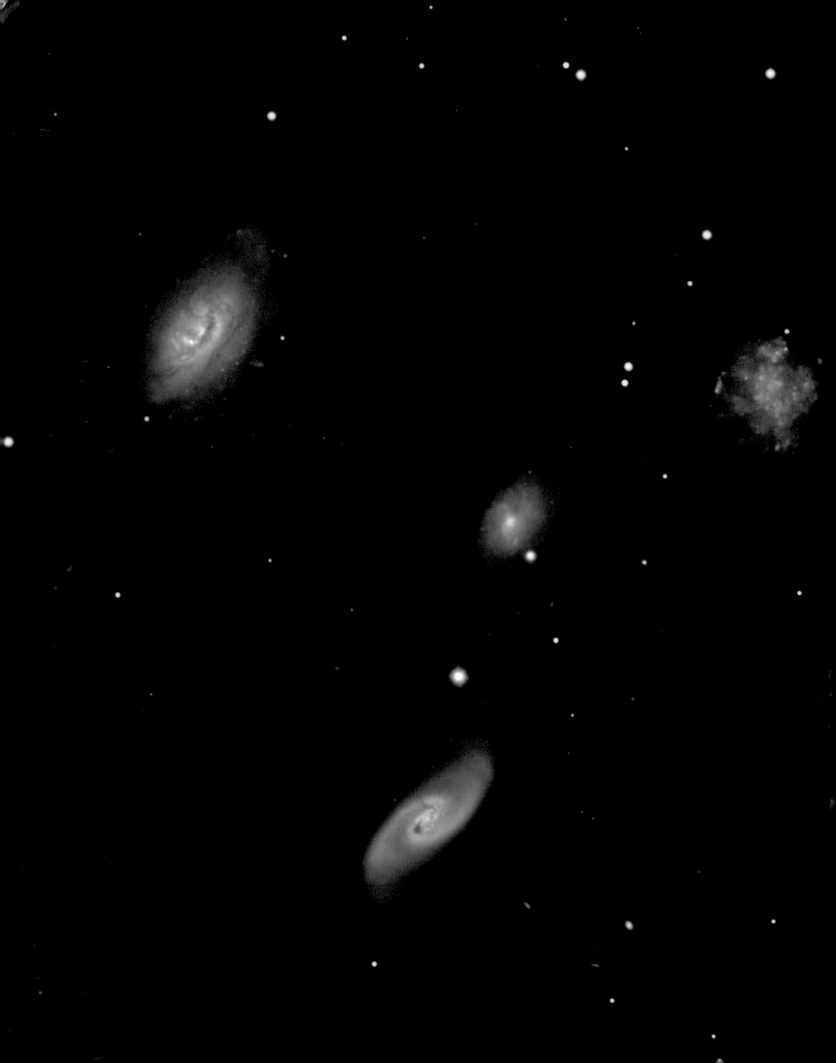

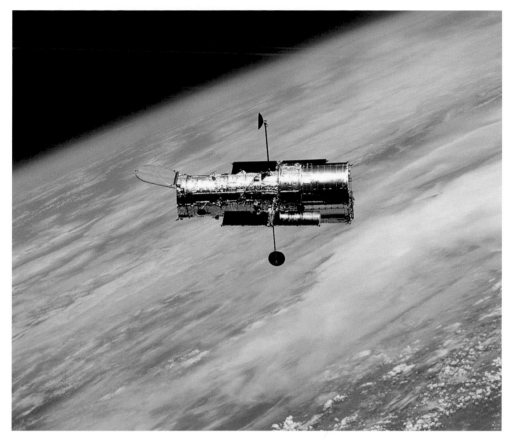

◀ **Hubble Space Telescope**
Barely skimming the Earth's atmosphere at a height
of 380 miles (600 km), this unique telescope has
utterly transformed our knowledge of the Universe
since its launch in 1990.

than red-hot ones, so the colours of the stars reflect their temperatures. Betelgeux, for
example, appears red because it is cooler than our own Sun, whereas Rigel – the other
bright star in Orion – is blue-white and is much hotter than our own, yellow Sun, which
is intermediate in temperature as well as in size.

The history of time

Because of the vast distances involved, when we look at the stars we are taking part in
time-travel, without any need for a Wellsian machine or Dr Who's Tardis. Consider Sirius,
which is the brightest star of the night sky and is very conspicuous for several months in
each year. It is 26 times as powerful as the Sun, and 8.6 light-years away – that is to say,
roughly 50 thousand billion miles. Its light takes 8.6 years to reach us, so that if we look
at it in the year 2007 we are actually seeing it as it used to be in 1999.

The Pole Star (Polaris), which many people (certainly all navigators) can recognize, is
about 400 light-years from us according to the latest measurements. The light we see
coming from it right now left Polaris around 1606 – and any astronomer there equipped
with a sufficiently powerful telescope could look at the Earth and see England as it used
to be in the time of Shakespeare.

The light now reaching us from Rigel started on its journey towards us in the time of
the Crusades, and even this is local by the standards of the Universe as a whole. We can
now study objects that are so remote that we see them as they used to be long before the
Earth existed.

◀◀ **Robert's Quartet**

This compact group of four galaxies is about 160
million light-years away. Apart from their beauty,
such small groups are excellent
laboratories to study galaxy interactions.

▶ Looking back to the Big Bang

Looking out into space, we are literally looking back in time. The light we see from the planets in our Solar System left them only minutes ago, but when we observe the most distant galaxies the Hubble Telescope has been able to photograph, we see them as they were 12 billion years ago. In this schematic diagram we, the observers, are at the bottom of the picture looking up. We will never be able to see light from the Big Bang itself, but we believe we know when it happened. On this diagram it provides the top, most-distant point – the beginning of time.

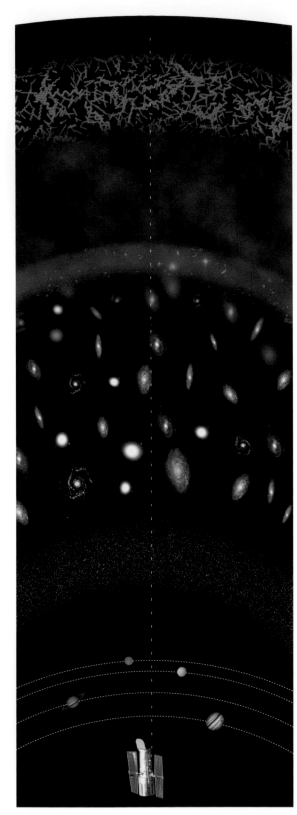

Big Bang, 13.7 billion years ago.

Opaque Universe, up to 13.4 billion years ago.

The Dark Ages.

First stars and galaxies formed 13.2 billion years ago.

Most distant observed galaxies. Light left them 12.7 billion years ago.

Many of the galaxies we see formed in this period.

Nearby galaxies, such as the Virgo cluster. Light left these galaxies 50 million years ago.

Nearest major galaxy Andromeda is about 2.5 million light-years away.

Nearest star, Proxima Centauri, is 4.3 light-years away.

We see Jupiter as it was roughly an hour ago.

You are here.

This time travel is essential for our understanding of the Universe; we can actually see much of the story we want to understand. For example, if we suspect galaxies were much smaller in the past, we can actually observe them to confirm this. By looking at galaxies six billion light-years away, we believe we are seeing the Universe that our own galaxy inhabited six billion years ago.

If the distance scale stretches the imagination, the timescale is equally staggering. Various lines of investigation tell us that the age of the Earth is approximately 4.6 billion years and that it was formed from a cloud of dust and gas surrounding the youthful Sun, but we humans are newcomers to the terrestrial scene. To drive this home, let us imagine a timescale in which the age of the Earth is represented by one year: it came into existence at midnight on January 1. Primitive life appeared by early May, but fish did not evolve until mid-November, and the first forays on to the land surface were achieved at the very end of November. Reptiles ruled the world during the first weeks of December; the dinosaurs died out around December 15 while mammals came unobtrusively into the picture, but only on the morning of December 31 did ape-men arrive. The whole story of *Homo sapiens* is compressed into the last hour of the last day of the year. Jesus Christ appeared on Earth less than a minute ago.

We are reasonably confident about our distance-scale and about the age of the Earth. We have also made tremendous progress in estimating the age of the Universe as we know it; the latest value of 13.7 billion years is probably accurate to within a few per cent. However, this introduces a really major problem.

The inescapable fact is that we exist; we are made up of atoms and molecules, and this material must have been created in one way or another. Either it has always existed, or else it was produced at a definite moment in time. Neither picture is easy to accept. If the material of which we are made has always existed, we have to visualize a period of time that had no beginning. If it came into existence suddenly, 13.7 billion years ago, what happened before that? *Was* there a 'before'?

The mathematical answer is that time began with the Universe, so that there was no 'before'. This may be theoretically accurate, but it is certainly unsatisfying. In studying the Universe we treat time as a fourth coordinate – we are sitting writing this at latitude 50 degrees north and longitude 0.41 degrees west a few metres above sea level, but in order to find us you also need to specify a time – we are to be found in late 2006.

Yet this simple picture breaks down over astronomical scales. Let us say that in the far future astronomers want to carry out an experiment simultaneously on Earth and on the nearest star, Proxima Centauri, which is located a little more than four light-years away. As no information can travel faster than light, a light signal sent between the two systems will not suffice to coordinate the experiment – time is not an absolute thing upon which all observers will agree.

In the uncertainties that we find surrounding us, we can only make intelligent guesses. This may sound haphazard, but this is essentially the scientific method. To explain an observed fact, a theory is put forward. The theory is then used to make predictions. By making new observations the predictions can be tested. If the predictions are confirmed, we have a good theory; if not, we must think again. In the following chapters, in constructing our model of the history of the Universe, we will be using the theories which have best stood up to the scrutiny of current experimental astronomy. And so to the beginning…

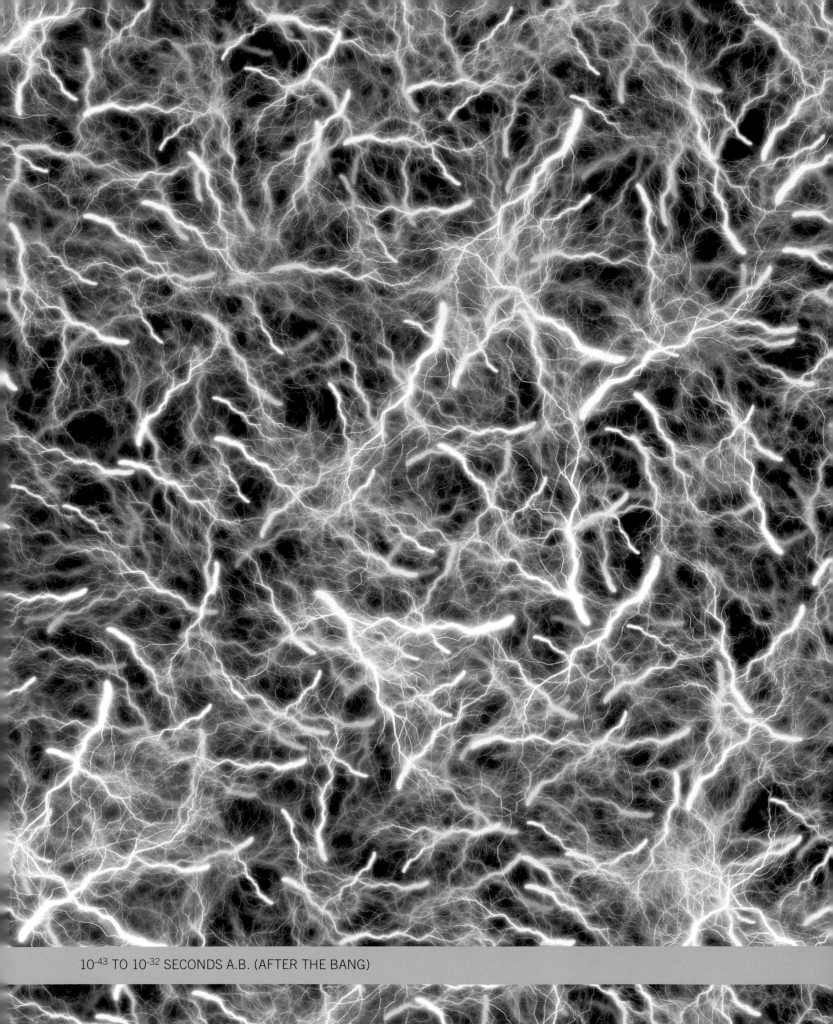

10^{-43} TO 10^{-32} SECONDS A.B. (AFTER THE BANG)

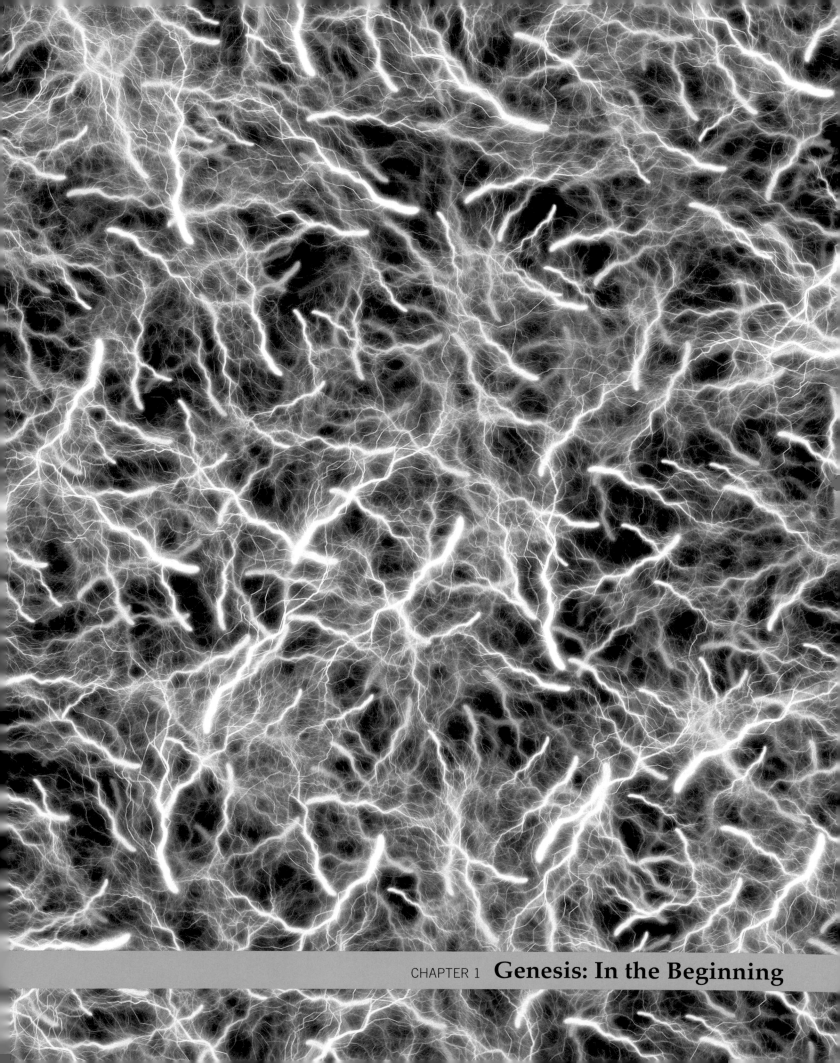

CHAPTER 1 **Genesis: In the Beginning**

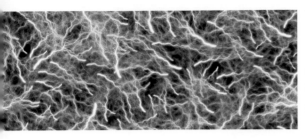

▲ Cosmic chaos

The chaotic, almost infinitely small Universe, a tiny fraction of a second after the Big Bang. In this representation the bright lines represent short-lived particles that are continually created and annihilated as they collide with each other.

▶▶ Andromeda Galaxy (M31)

Looking out through the foreground stars of our Galaxy, we see our nearest large neighbour in its entirety, comprising perhaps 200 billion stars. At a distance of 2.5 million light-years, it is one of the most distant objects visible to the naked human eye.

Everything, space, time and matter, came into existence with a 'Big Bang', around 13.7 billion years ago. The Universe then was a strange place – as alien as it could possibly be. There were no planets, stars or galaxies; there was only a melée of elementary particles; these filled the Universe. Moreover, the entire Universe was smaller than a pin-prick, and it was unbelievably hot. At once it began to expand – and as it spread out from this bizarre, unexpected start, it evolved into the Universe we see today.

Modern science is unable to describe or explain anything that happened in the first 10^{-43} seconds after the Big Bang. This interval, 10^{-43} seconds, is known as the Planck time, named after the German scientist Max Karl Ernst Planck. He was the first to introduce the concept that energy could be regarded not as a continuous flow, but as packets, or 'quanta', each with a specific energy. Quantum theory is now at the base of much of modern physics; it deals with the Universe on the smallest scales and is certainly one of the two great achievements of 20th-century theoretical science. The other is Einstein's general theory of relativity, which deals with the physics of very large scales – astronomical scales, in fact.

Despite the fact that, in their own realms, both theories are extremely well tested by experiments and observations, there are major problems in reconciling these theories with each other. In particular, they treat time in fundamentally different ways. Einstein's theories treat time as a coordinate; it is therefore continuous, and we move smoothly from one moment to the next. In quantum theory, however, the Planck time represents a fundamental limit – the smallest unit of time that can be said to have any meaning at all, and the smallest unit that could ever, even in theory, be measured. Even if we built the most accurate clock possible, we would see it jump rather erratically from one Planck time to the next.

Attempting to reconcile these two contrasting views of time is one of the major challenges for 21st-century physics ('string theory' and its cousin 'membrane theory' are currently closing in on this goal, but there is much work still to be done). For now, in the small, hot, dense Universe that existed just after the Big Bang, quantum physics holds sway, and hence we begin our scientific study of the Universe 10^{-43} seconds after the beginning.

The Big Bang is a counter-intuitive idea. Our common sense seems much more impressed by a static and infinite Universe, and yet there are good scientific reasons to believe in this singular event. If we accept the Big Bang, it is possible to trace the whole sequence of events from that first Planck time right up until the present day, where we find ourselves on what Carl Sagan memorably described as our 'pale blue dot'.

The beginning of time

So let us look back to the very start of the Universe – just after the Big Bang itself. It is tempting to picture the Universe suddenly bursting out in a vast ocean of space, but this is completely misleading. The true picture of the Big Bang is one in which space, matter and, crucially, time were born. Space did not appear out of 'nothingness'; before the moment of

Standard form

10^{-43} is a convenient form of mathematical shorthand, and could be written as a decimal point followed by 42 zeros and then a 1. This is precise enough, but is decidedly clumsy (0.000000000000 000000000000000000000000000001). In order to deal with the incredibly large and incredibly small numbers that pervade astronomy, we will use so-called standard form throughout the book. 10^{33}, for example, is 1 followed by 33 zeros – we could write 1000 as 10^3, or 0.001 as 10^{-3}.

▶▶ Large-scale structure in the Universe

This enormous cluster of galaxies (AC03627), 250 million light-years away, is typical of what we might see in every direction, if we could see past the dust and gas of our own Galaxy and its neighbours. Clusters such as these are the largest objects in the Universe to be held together by their mutual gravitational pull.

▼ Cubic Space Division

Dutch artist M. C. Escher created this lithograph in 1952. Born in Holland in 1898, his work became internationally recognized after his first important exhibition was reviewed by *Time* magazine in 1956. Mathematicians recognize the extraordinary visualization of their abstract principles in his work.

creation there was no 'nothingness'. Time itself had not yet begun, and so it does not even make sense to speak of a time before the Big Bang. Not even a Shakespeare or an Einstein could explain this in plain English, though the combination of the two might be useful!

It also follows that when we survey the Universe today, it is meaningless to ask just 'where' the Big Bang happened. Space only came into existence with the Big Bang itself. Hence, in those first few fractions of a second, the entire Universe we see today was in a tiny region, smaller than an atomic nucleus. The Big Bang happened 'everywhere', and there was no central point.

A nice illustration of this is given in a famous painting by Escher, known somewhat unromantically as *Cubic Space Division*. Imagine standing on any of the cubes that mark junctions on this lattice, while each and every one of the rods joining the cubes expands. From your perspective, it would seem that everything is rushing away from you, and it might seem natural at first to conclude that you are in a special location – the centre of the expansion. Yet standing back and thinking allows you to realize that the expansion would look the same wherever in the lattice you were; there is no centre. The situation is very similar in our Universe; each group of galaxies appears to be rushing away from us, and yet observers looking back at us from these distant stars would see the same illusion and would presumably be just as likely to conclude that they are all at the centre of the expansion.

Another problem concerns the often-asked, and at first glance, sensible question, 'how big is the Universe?' Here we again have a major problem – there seem to be two possible answers. Either the Universe is of finite size, or else it isn't. If finite, what lies outside it? The question is meaningless – space itself exists only within the Universe, and therefore there is literally no 'outside'. On the other hand, to say the Universe is infinite is really to say that

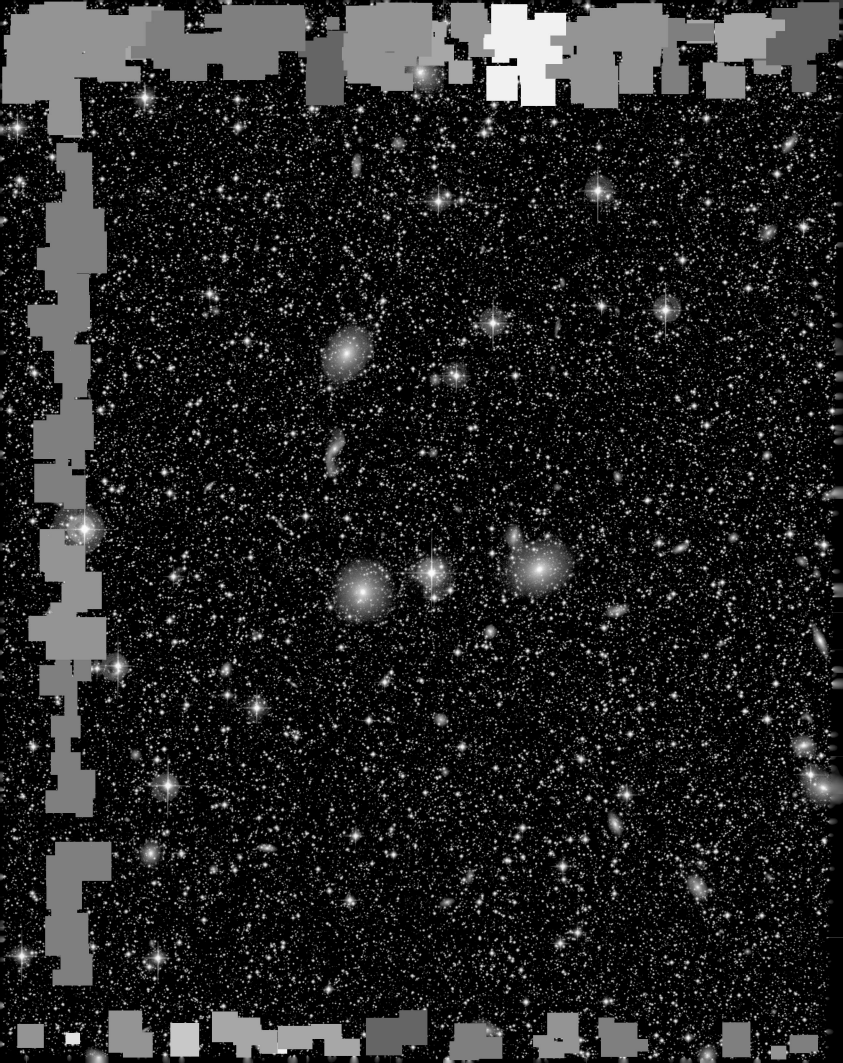

▼ From the invisible to infinity

We can now study objects at both extremes in scale. The size of human beings is between one and two x 10^0 (i.e. one) metres. If we measure the size of the Earth, it comes in at a few x 10^6 (i.e. a few million) metres. Our range of experience ranges from about 10^{-15} metres, the scale of fundamental particles that make up atoms – quarks, etc, to 10^{25} metres, the scale of the entire observable Universe.

its size is not definable. We cannot explain infinity in everyday language, and neither could Albert Einstein (we know, because Patrick asked him!).

Remember, too, that we need to consider time as a coordinate; in other words, we cannot simply ask 'how big is the Universe?' as the answer will change over time. We could ask 'how big is the Universe now?', but as we shall see later, a consequence of relativity is that it is impossible to define a single moment called 'now' that has the same meaning across the entire Universe.

Talking about a Universe that has a particular size immediately leads to thoughts of an edge. If we travelled far enough, would we hit a brick wall? The answer is no – the Universe is what mathematicians call finite but unbounded. A useful analogy is that of an ant crawling on a ball. By travelling always in the same direction on its curved surface it will never come up against a barrier, and it can cover an infinite distance. This is despite the finite size of the ball, to which the ant will be completely oblivious. Similarly, if we were to set off in a powerful spacecraft in what we perceive to be a straight line, we would never reach the edge of the Universe – but this does not mean that the Universe is infinite; we shall see later that space, too, may be regarded as curved.

So let us restrict ourselves to questions we can answer scientifically, meaning questions we can answer through comparison with observation. We can say for certain that the

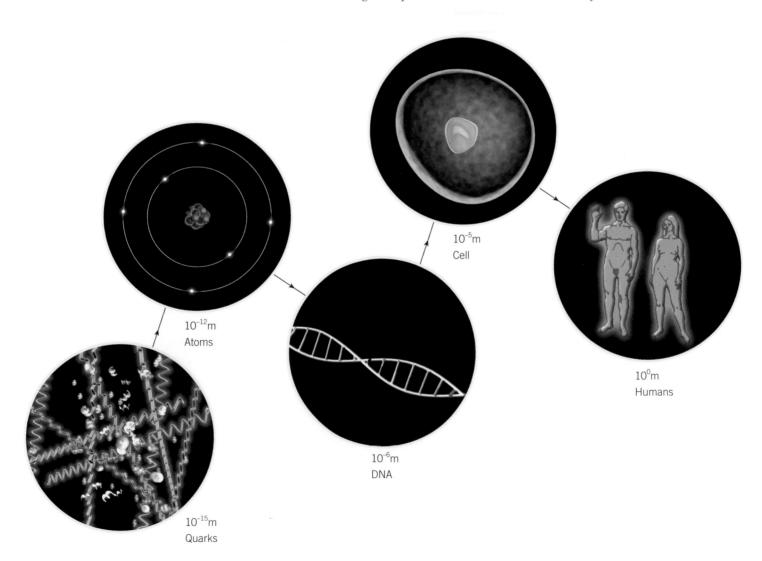

10^{-5}m
Cell

10^{-12}m
Atoms

10^0m
Humans

10^{-6}m
DNA

10^{-15}m
Quarks

observable Universe (literally that part of the Universe from which light can potentially have reached us) is finite in size because, at our current best guess, the Universe is only 13.7 billion years old. Therefore the edge of the observable Universe, from where light could only just be reaching us, must be 13.7 billion light-years away, and expanding at a rate of one light-year per year. In fact there are reasons why we will never be able to see quite this far, as will become clear later. All we can say for certain about the size of the Universe is that it must be larger than the portion we can see.

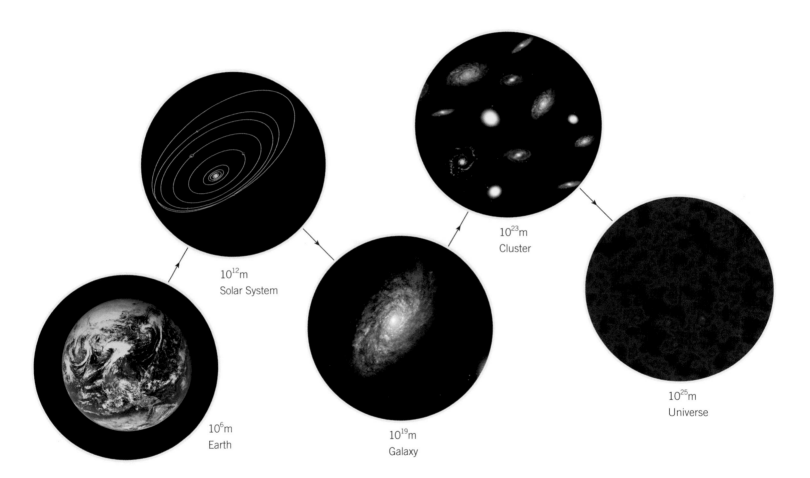

10^{12}m
Solar System

10^{23}m
Cluster

10^{6}m
Earth

10^{19}m
Galaxy

10^{25}m
Universe

The scale of the Universe

Of course, saying that an object is 13.7 billion light-years away is all very well, but can we really comprehend the scale of the Universe? It is possible to appreciate fully the distance between, say, London and New York, or even the distance between the Earth and the Moon – roughly a quarter of a million miles – which is roughly ten times the circumference of the Earth, and many people have flown for a greater distance than that over their lives. Indeed, several airlines give special privileges to those who have travelled more than a million miles in their lifetimes. But how do you really grasp 93 million miles, the distance of the Sun? And when we consider the nearest star, 4.2 light-years (approximately 24 million million, or trillion, miles) away, we are quite unequal to the task…. The galaxies are immensely more distant than this: even the Milky Way's nearest neighbours, such as the Andromeda Galaxy, are over two million light-years away.

▲ Atomic layers

A microscope image of atomic layers on the surface of a crystal of iron silicide. The smallest step is only one atom thick.

At the other end of the scale, visualizing the size of an atom, which cannot be seen individually with any ordinary microscope, is equally difficult. It has been said that, in scale, a human being is about half-way between an atom and a star. Interestingly, this is also the regime in which physics becomes most complicated; on the atomic scale, we have quantum physics, on the large scale, relativity. It is in between these extremes where our lack of understanding of how to combine these theories really becomes apparent. The Oxford scientist Roger Penrose has written convincingly of his belief that whatever it is that we are missing from our understanding of fundamental physics is also missing from our understanding of consciousness. These ideas are important when one considers what have become known as anthropic points of view, best summarized as the belief that the Universe must be the way it is in order to allow us to be here to observe it.

Another useful question to ask is how many atoms there are in the Universe. One estimate has come up with a total number as large as 10^{79}, or in other words a 1 followed by 79 zeros.

Traditionally we have viewed atoms as being made up of three more fundamental particles: the proton (carrying unit positive electric charge), the neutron (no charge at all) and the much less massive electron (carrying unit negative charge). Incidentally, it is far from easy to define what electric charge is on the atomic level. It will suffice to think of charge as a property that particles can have, just as they have a size and a mass. Charge always comes in parcels of fixed size which we call unit charge.

Classically these particles are considered to be arranged like a miniature solar system, with electrons orbiting a central compound nucleus, containing protons and neutrons. This nucleus carries a positive charge, which is exactly balanced by the combined charge of the orbiting electrons. In our Solar System of planets, the force of gravity keeps the planets in their orbits around the central Sun, but here in the atom it is the attraction between the negatively charged electron and positive nucleus that keeps the electrons in their orbits.

In passing, we should note that this simple picture can explain much of basic chemistry; for example, why it is the outer electrons of atoms that tend to be involved in chemical reactions. They are further away from the nucleus, and therefore they are less tightly held by its attractive force. So the simplest atom, that of hydrogen, has a single proton as its nucleus, and one orbiting electron. The whole atom is therefore electrically neutral: plus one added to minus one equals zero. All atoms have an equal number of electrons and protons. Each element has a unique number of these particles, known as the atomic number. For example, helium atoms have two protons and electrons – an atomic number of two – while carbon atoms have an atomic number of six. Heavy elements have large numbers of protons and electrons. Uranium, the heaviest natural element on Earth, has an atomic number of 92.

This view of the atom, which saw protons and neutrons as solid lumps, prevailed in the early 20th century, but things are much less clear-cut today. Much of the strange behaviour of extremely small systems can only now be explained by considering them to be made up of waves rather than particles. This theory is known as wave-particle duality. In addition, experiments have shown that while electrons seem to be truly unbreakable, protons and neutrons are not in fact fundamental – they can be split into smaller particles known as quarks, which themselves are now believed to be fundamental. No one has ever seen a quark, but we know they must exist since they have been detected in particle accelerators that have been built to smash protons together at incredibly high speeds. In these experiments protons are seen to fracture, and hence scientists conclude that they cannot be fundamental. Nature abhors a naked quark; quarks appear only in pairs or triplets.

The forces of nature

The reason for this property of quarks lies with an unusual property of the force that normally binds quarks together, known (not without reason) as the strong nuclear force. It is dominant on very small scales, which is why we need such powerful particle accelerators to smash protons apart. Unlike the forces with which we are familiar on larger scales, such as gravity or the attraction between opposite electric charges, the strong force increases with distance. In other words, if we could separate two quarks we would find them being pulled together more and more strongly as the distance between them increases. Eventually, as quarks move apart from each other the energy caught up in straining to pull them apart becomes so great that two extra quarks are produced, the energy being converted into mass. Suddenly we have two pairs of quarks rather than the individual quark we were attempting to isolate. This process means that no experiment ever produces individual quarks, and in the everyday Universe they exist only as components of other particles, such as protons and neutrons, which each contain three quarks.

At the huge temperatures in the Universe immediately after the Big Bang, the quarks had enough energy to roam free, and so by understanding the story of the Universe on the largest scales we may come to understand more about the particles that account for the smallest scales. The energy that each particle possessed in the early Universe will remain

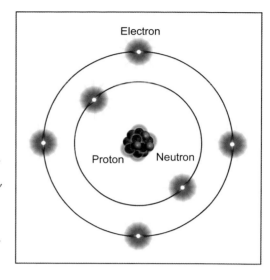

▲ The classical atom

This simple model of an atom was proposed by Niels Bohr in 1913. It involves a nucleus made up of neutrons (blue) and protons (red) that is surrounded by electrons orbiting like the planets around the Sun. Despite the advent of quantum mechanics, which paints a very different picture of the atom and shows fuzzy 'probability densities' in place of the ball-like particles, Bohr's classical model is still useful today.

▶ **Hunting quarks**

At Brookhaven National Laboratory in New York, beams of gold nuclei are smashed together at close to the speed of light. The result is a recreation of a state of matter that is believed to have existed ten millionths of a second after the Big Bang, known as a 'quark-gluon plasma'. This image, looking remarkably like a human eye, shows the tracks of about 1000 particles emerging from a collision. It is effectively a cross-section; the particles are actually sent off in all directions.

far beyond the reach of our particle accelerators; even an accelerator the size of the Solar System would be incapable of producing particles with this enormous energy.

It is a remarkable fact that our present day research into the very small, via particle physics, and into the very largest scales, via cosmology, are so intertwined. To understand the whole Universe, we are dependent on our understanding of the fundamental particles, and the best tests for our theories about them are in the embryonic Universe. A 'hot space' full of these highly energetic fundamental particles is the earliest image we can conjure of our newly-born Universe.

Bigger is cooler

From the first Planck time onwards, this inconceivably small, inconceivably hot Universe began to expand and hence also to cool. The Universe was a sizzling ocean of quarks, each of which had a vast amount of energy, moving at a huge speed. As a result, there could be no atoms or molecules of the kind we know today, because these are complicated structures, quite unable to survive the disruption of very high temperatures; the quarks were simply too energetic to be captured and confined within protons or neutrons. Instead, they were free to career around in the infant Universe, until they collided with their neighbours. As well as quarks, this early soup of sub-atomic particles also contained antiquarks – identical twins, but bearing opposite electric charges. It is now believed that each particle has its equivalent antiparticle, identical in all respects other than its electric charge, which is opposite. The antimatter particle corresponding to an electron is a positron, which bears a positive charge, but is otherwise exactly similar to the electron. The concept of antimatter is familiar from

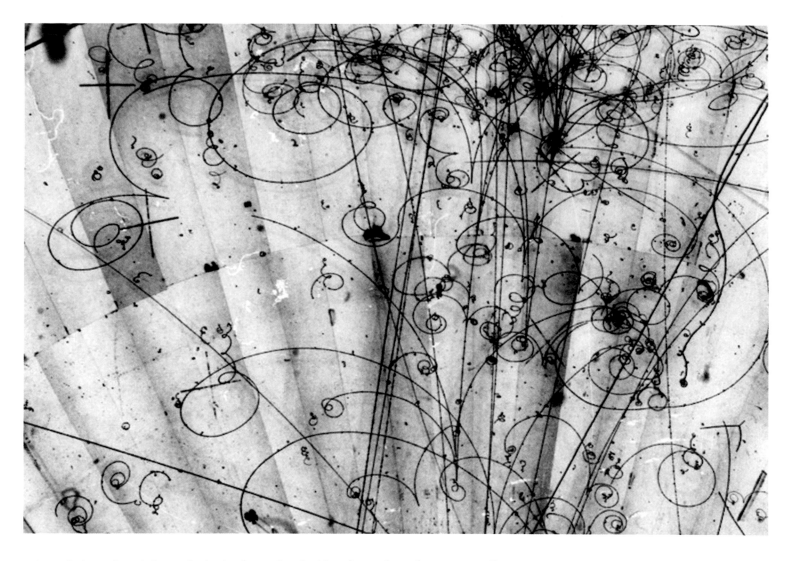

science fiction, where it forms the basis of countless highly-advanced starship engines, all of which are based on the fact that a collision between a particle and antiparticle results in the annihilation of both and the release of much energy – this has been verified by experiments. Whenever a quark met an antiquark in the primeval Universe both would vanish, releasing a flash of radiation. The reverse process also occurred; radiation of sufficiently high energy (certainly at the energies found at this early stage of the evolution of the Universe) could spontaneously produce pairs of particles, each pair composed of a particle and its antiparticle. The Universe at this epoch, then, was composed entirely of radiation that produced pairs of particles, which in turn vanished as they collided with each other, returning their energy to the background radiation.

As the Universe continued to expand and cool, after the first microsecond (only ten million million million million million million Planck times), when the temperature dropped below a critical value of about ten million million degrees, the quarks slowed down enough to enable them to be captured by their mutual (strong force) attraction. Bunches, each of three quarks, clumped together to form our familiar protons and neutrons (collectively known as baryons), whereas the antiquarks clumped together to form antiprotons and antineutrons (antibaryons). Had the number of baryons and antibaryons

▲ **Seeing antimatter**

This image shows the creation of pairs of electrons and positrons in a bubble chamber. Charged particles leave a trail of minute bubbles behind them, allowing the eye – or the camera – to track their progress. The photons that provide the energy to create the electrons and positrons cannot be seen, as they have no charge. Each pair of trails begins at a common point, from which the particles can be seen to spiral outwards. Moving in a strong magnetic field, the electrons and positrons experience opposite forces, and so spiral in opposite directions.

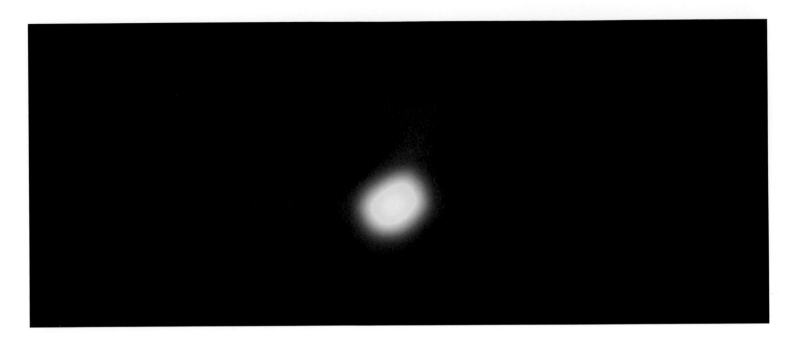

▲ Cloud of antimatter

This gamma-ray map of the centre of our Galaxy
is believed to show matter and antimatter particles
– electrons and positrons – colliding with each other,
producing the annihilation of both sets of particles, and
the emission of huge amounts of energy. This implies
that antimatter in the form of positrons is streaming
from the centre of our Galaxy.

been equal, the most likely outcome is that collisions between them would have resulted in
complete annihilation. The energy left in the radiation would have become diluted as the
Universe expanded, and so new particle pairs would no longer have been created. Matter in
the Universe would not have survived to the present day.

It was only the fact that there was a very slight imbalance built in from the very start
that saved matter and therefore enables us to be here, wondering what happened in these
remote times. Due to reasons we do not yet understand, for every billion antibaryons there
were a billion *and one* baryons, so that when the grand shoot-out was over, almost all the
antibaryons had vanished, leaving behind the residue of protons and neutrons which make
up the atomic nuclei of today.

The cosmic conspiracy

Let us return briefly to the present; consider two galaxies, each nine billion light-years
away from us, but in opposite directions as seen from the Earth. The distance between
them is therefore 18 billion light-years. Both will exist in regions of the Universe that,
broadly speaking, look the same on the largest scales. One may be deep in the heart of a
cluster of galaxies, such as our own nearby Virgo Cluster, whereas the other may be much
more isolated. And yet near the first there will be isolated galaxies, and near the second
galaxy there will inevitably be a galaxy cluster. Each region will contain the same types of
galaxies in the same proportion, and even the temperature of the local region will be the
same in both cases.

This constitutes a problem known as the 'cosmic conspiracy'. The Universe is less than
18 billion years old (remember the best estimate is 13.7 thousand million years), so light
would not yet have had enough time to make the crossing between the two galaxies, and
relativity insists that light is the fastest thing in the Universe. If light has not yet had time
to cross the space between the two regions, then nothing else could have done, and so
nothing could have passed from the first region to the second. Any differences between the
regions could not be ironed out, and it is therefore surprising that the Universe seems much

the same in whichever direction we look; we see galaxies of the same type, distributed in much the same way, and it is this which is called the 'cosmic conspiracy'.

Why is this a problem? Doesn't it seem natural that the Universe appears the same in whichever direction we look? Perhaps there is some as-yet-unknown law, governing the physics of the Big Bang itself, which ensures that only universes that are almost completely uniform can be produced. However, we have no hint of any physics that could predict this, and so we must at least consider the possibility that the Universe began with large differences in temperature between different regions, for example, an early Universe in which one half is twice as hot as the other half. How could this lead to the kind of uniformity we see today? There has not been time for heat to flow from the warm to the cool region of the Universe, and there has not even been time to send a message, travelling at the speed of light. In such circumstances, correcting this original imbalance would seem to be impossible and yet these widely separated, disconnected areas are, in fact, similar.

Our two galaxies may be far apart now, but when the Universe was very young it was also much smaller and bodies on opposite sides could have been in touch and able to exchange heat, producing the uniformity seen today. The question now, therefore, is how big the Universe was in these early stages. Surprisingly this seems to be a fairly simple question to answer.

Only one of the forces that we have so far discussed can act over astronomical distances and that is gravity, which by its very nature is an attractive force, pulling material together. Gravity alone would slow down the expansion from its initial rate. We can attempt to work backwards from the present day to determine how the size of the Universe has changed with time – and we discover that the cosmic conspiracy survives into the early Universe. In other words, the Universe was *never* small enough to allow light to cross from one side to the other, and therefore never small enough to allow temperature differences to even out. This whole scenario presupposes that gravity is the only force affecting the rate of expansion, and it is only if we are prepared to abandon this idea that we can solve the conspiracy problem.

The crazy edifice of inflation

The currently popular solution involves increasing the complexity of the Big Bang theory somewhat. Most cosmologists now believe that there was an extremely short period of rapid expansion, known as inflation, between 10^{-35} and 10^{-32} seconds after the Big Bang,

▼ **The cosmic conspiracy**

While we can see remote galaxies A and B in opposite directions in the sky, they cannot see each other. In the whole of the time since the Big Bang, light has still not had time to travel between them.

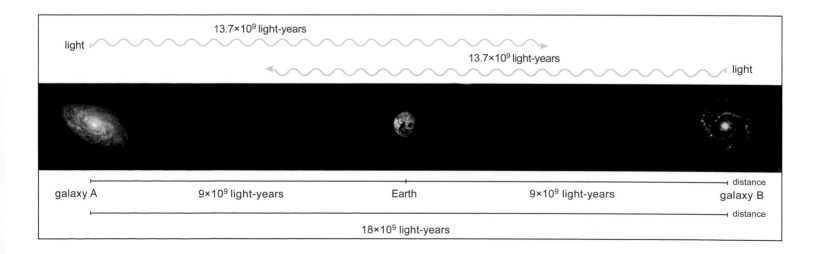

13.7×10⁹ light-years

light

13.7×10⁹ light-years

light

galaxy A   9×10⁹ light-years   Earth   9×10⁹ light-years   galaxy B

distance

distance

18×10⁹ light-years

▲ Hubble Deep Field South

Galaxies seen at a distance of 12 billion light-years. The distance between the most remote galaxies seen in the Hubble North and South Deep Fields is so great that in the 13.7 billion year history of the Universe, light would not have had time to get much beyond halfway between them, and yet they look similar, as does the Universe they find themselves in. A striking example of the cosmic conspiracy.

during which period the size of the Universe increased many billion times. At the end of the inflationary period, the expansion settled back to a relatively stable rate, consistent with that observed today.

Without a period of inflation, regions of the Universe we see on opposite sides of the sky would neither have had time to exchange heat nor to settle down into a comfortable equilibrium. The suggested rapid expansion allows us to believe that the Universe was initially much smaller, and could therefore reach a uniform temperature before the acceleration began. Any remaining small differences would then be ironed out by the vast increase in scale. This is because another consequence of the amazingly rapid inflation is that the region of the Universe we observe is only a tiny fraction of the entire Universe. In other words, we are only looking at variations in what is effectively our local neighbourhood and these are bound to be small. To give an analogy much nearer home, the Earth as we see it is marked by huge variations in height; from the peak of Mount Everest to the bottom of the deepest ocean trench. Inflation achieves the equivalent effect of blowing up the patch of ground under your little toe to the size of the entire globe (or equivalently, shrinking us to a size much smaller than the smallest virus). The differences in height that we can reach and explore are then bound to be slight; inflation has exactly the same influence on temperature fluctuations in the Universe.

But why should the infant Universe suddenly undergo such an extreme increase in the speed of its expansion? It seems that there is a need to introduce a new force, which can be held responsible for the vast acceleration, acting in the opposite direction to gravity. Scientists have begun to study in detail what properties such a force might have and yet there seems to be no obvious explanation. As far as we know, there are no particular circumstances peculiar to the epoch just prior to inflation, and the appearance and sudden disappearance of this accelerating force therefore seems to be somewhat arbitrary, but it does at least allow us to deal with the problem of the 'cosmic conspiracy'.

Are there other problems which the introduction of inflation can solve for us? It turns out that inflation can also explain two other features of the Universe we see today, which are otherwise completely inexplicable. First, according to the standard theory of particle physics, a certain type of particle, known as a monopole, should occasionally appear in detectors. In fact, none has ever been detected, and this requires some explanation. The theory of inflation allows us to argue that the concentration of these particles has become so low that our failure to locate any is not surprising. Say, for the sake of argument, that 100 million million of these particles were created in the Big Bang; it would be surprising that we have failed to detect a single one. But if the same number of monopoles were created and spread through a universe that has become many thousands of millions of times larger than before inflation, it is likely that there could well be none within the entire visible Universe. The speed of inflation was so overwhelming that, even in the short time for which it operated, it produced a universe which was inconceivably larger than that predicted by the conventional Big Bang. Inflation provides an explanation for the missing particles – they have simply been diluted away.

Life in a flat Universe

There is a third pillar supporting the seemingly crazy edifice of inflation, perhaps the most convincing of all. This involves the geometry of the Universe. Most people are familiar with the geometry of Euclid which we learn, sometimes perhaps reluctantly, at school. We are told the three angles of a triangle add up to 180 degrees. However, this is not always the

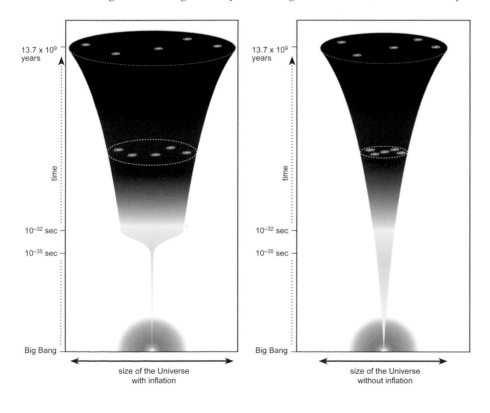

13.7 x 10⁹ years

10⁻³² sec

10⁻³⁵ sec

Big Bang

size of the Universe with inflation

size of the Universe without inflation

◀ Inflation

The left-hand diagram shows the impact of inflation compared with the Universe that did not undergo inflation (right). As the Universe expands, the Galaxies move further apart. With inflation, the Universe is smaller just after the Big Bang, but much larger today.

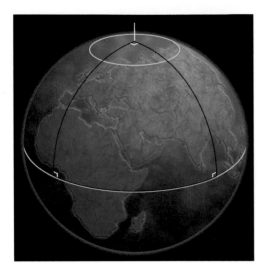

▲ Spherical geometry

In the Euclidean geometry we learn in school, the interior angles of a triangle always add up to 180 degrees. But a triangle drawn on a sphere can contain more than 180 degrees. This triangle drawn on the Earth's surface contains three right angles, adding up to 270 degrees.

case; on the surface of a sphere, for example, they can add up to more than 180 degrees. Consider a line drawn from the north pole, down the Greenwich meridian to the equator, and then east along the equator, a turn of 90 degrees. If we complete the triangle by returning to the pole along the meridian which runs through Russia, we will make another 90 degree turn. 90 + 90 =180, and we still have to add in the angle at the top, between the two meridians. Euclidian geometry applies only to flat surfaces.

What form does the geometry of the Universe take? Things are complicated because we are dealing with a four-dimensional space (the three familiar spatial dimensions, plus time) rather than a two-dimensional surface as before. Let us consider the largest scales, and ignore local distortions caused by matter. There are huge numbers of possible geometries for the Universe, and yet it seems that our Universe has been finely tuned to select one specific type; observations show (see Cosmic Microwave Background radiation in Chapter 2) that we live in what is known as a 'flat' universe, one in which Euclidian geometry holds even on the largest scales. Why should this be so? To achieve a flat universe, we need to have exactly the correct amount of matter within the Universe, to within a few atoms. In other words, if there were a few atoms too many or too few we would be in a universe with a geometry far from flat.

Once again, we are confronted with an observation that might be ascribed to some feature of the early physics governing the Big Bang itself – and once again inflation allows us an alternative and more satisfying explanation. The argument rests, as before, on the fact that inflation provides a much larger universe than the simple Big Bang.

Let us consider a three-dimensional analogy to help us think in four dimensions. Anyone standing on top of a bowling ball will quickly discover that it is a sphere, presumably when he or she falls off. What about a larger sphere, such as the Earth, upon which we happily stand every day? It may not be immediately obvious that we are standing on a curved surface, although it is certainly easy enough to work out. Contrary to popular belief, as long ago as the time of the ancient Greeks it was known that the Earth is a sphere (they even succeeded in measuring its diameter) and just watching a ship disappear over the horizon provides a clue that the Earth's surface is curved. Now imagine being on the surface of a sphere several trillion times larger than the Earth. All our experiments would indicate that we really were on a flat surface; the curvature due to the sphere would simply be too small to be detected – our ship would take an impossibly long time to reach the horizon.

Temperature

In everyday life we mostly measure temperature in degrees Celsius (Centigrade), whose scale is based around the freezing and boiling points of water. But temperature in the scientific sense is defined differently. It is based upon the speeds at which the atoms and molecules are moving around – the faster the speeds, the higher the temperature.

Holding a thermometer in liquid, what one is really measuring is how hard the molecules of the liquid are smashing into the thermometer, and hence how fast they are moving. This definition of temperature leads to some counterintuitive results:

consider a firework sparkler and a glowing poker. Each spark is white hot, but contains so little mass that it is quite safe to hold the handle of the firework – but most people would be very reluctant to grasp a red-hot poker, even though it is much cooler than the sparks of the firework.

Another, beautiful example is the pearly white corona, the Sun's outer atmosphere, which becomes visible during a total eclipse. The visible surface of the Sun is at a temperature of around 6000 °C, while the corona (for reasons not well understood) is at around a million degrees. Yet it is

After inflation

A Universe that has undergone inflation is like this last sphere. Because it is inflated to such an immense size, our observable Universe is only a tiny proportion of it and we can measure only its local properties. We may well conclude, rightly, that the Universe we can see has a flat geometry. In such a vast universe we can know nothing about the geometry of the Universe beyond the range of our observations. Regardless of which of the many possible geometries the Universe has, inflation tells us why our measurements indicate it is flat.

These three problems are all neatly solved by the idea of inflation, although at the price of introducing this mysterious, temporary acceleration, which remains poorly understood. Perhaps when we come to a greater understanding of the Big Bang itself then we will have an alternative answer, but for now inflation seems to be as good an explanation as any.

After the end of inflation, the Universe continued expanding and cooling at a lesser rate. Around three seconds after the Big Bang, the temperature had dropped to about a billion Kelvin. About three-quarters of the material in the Universe was hydrogen, and almost all the rest was helium (remember that the helium atom has two electrons, orbiting a nucleus composed of two protons and two neutrons).

The Big Bang predicts that for every ten protons, or hydrogen nuclei, produced there was one helium nucleus. Today hydrogen atoms still outnumber helium by ten to one. This provides perhaps the most simple and yet powerful test of the Big Bang theory. Stars convert hydrogen into helium, and so we would only expect the ratio to shift in favour of helium. If we observed a single object, anywhere in the Universe, in which there was less helium than expected we would have to completely reconsider our theory. No such observation has ever been made.

So do we believe in the Big Bang? Its major competitor, the steady-state theory, now seems finally dead. For now, the Big Bang holds the stage alone. We must remember that it is impossible to prove a theory, and all one can hope to do is ensure it is consistent with all the available evidence. The Big Bang with inflation appears to satisfy this requirement. However, at any moment a new discovery could expose a fatal flaw in the theory. Until a new Newton or another Einstein conjures up something better, we must live with the Big Bang.

▼ Total solar eclipse – double exposure

A total eclipse of the Sun is the grandest spectacle nature has to offer, and because of the huge range of brightness present at any one instant, no single photograph has the dynamic range to do it justice. Brian took this image of his first total eclipse in 1991 at Cabo, St Lucas, in Baja California, Mexico. By chance the filter used for the total phase was tilted at an angle to the camera lens, and by re-reflection produced a ghost image, clearly showing two prominences, which were massively overexposed in the main image – an exposure well-judged to show the beautiful shape of the fainter outer corona.

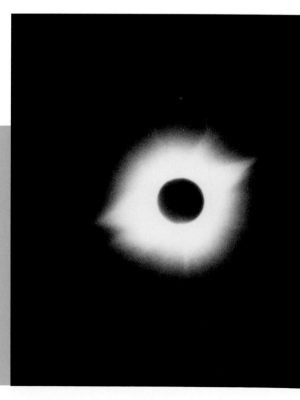

so tenuous that it would be perfectly possible to fly a spacecraft through the corona and survive. There would simply not be enough hot matter to significantly heat the spaceship.

The slower the movements of the atoms, the lower the temperature. Go down to around –273 °C (–472 °F) and the movements would have stopped completely. The temperature cannot fall further; we have reached absolute zero, the coolest possible temperature. This has never been (and never will be) attained in the laboratory, though we have worked down to a tiny fraction of a degree above

it, where matter takes on some extremely strange properties.

The Kelvin scale, named after Lord Kelvin, begins at absolute zero (0 K), but its degrees are the same size as those of the Celsius scale. To convert Kelvin to Celsius, subtract 273; to convert Celsius to Kelvin add 273. So 3 K is equal to –270 °C.

The advantage of the Kelvin scale is that we do not have to deal with negative numbers, and its zero point remains fixed and does not depend on pressure, unlike the boiling point and melting point of water.

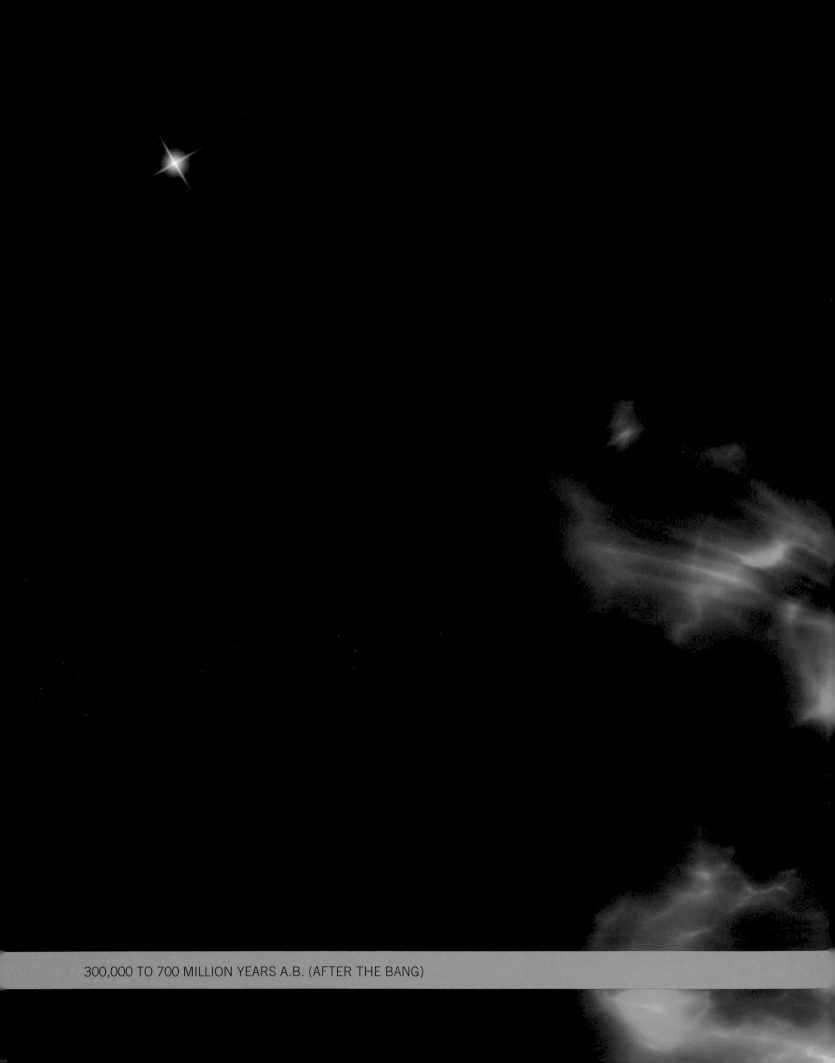

300,000 TO 700 MILLION YEARS A.B. (AFTER THE BANG)

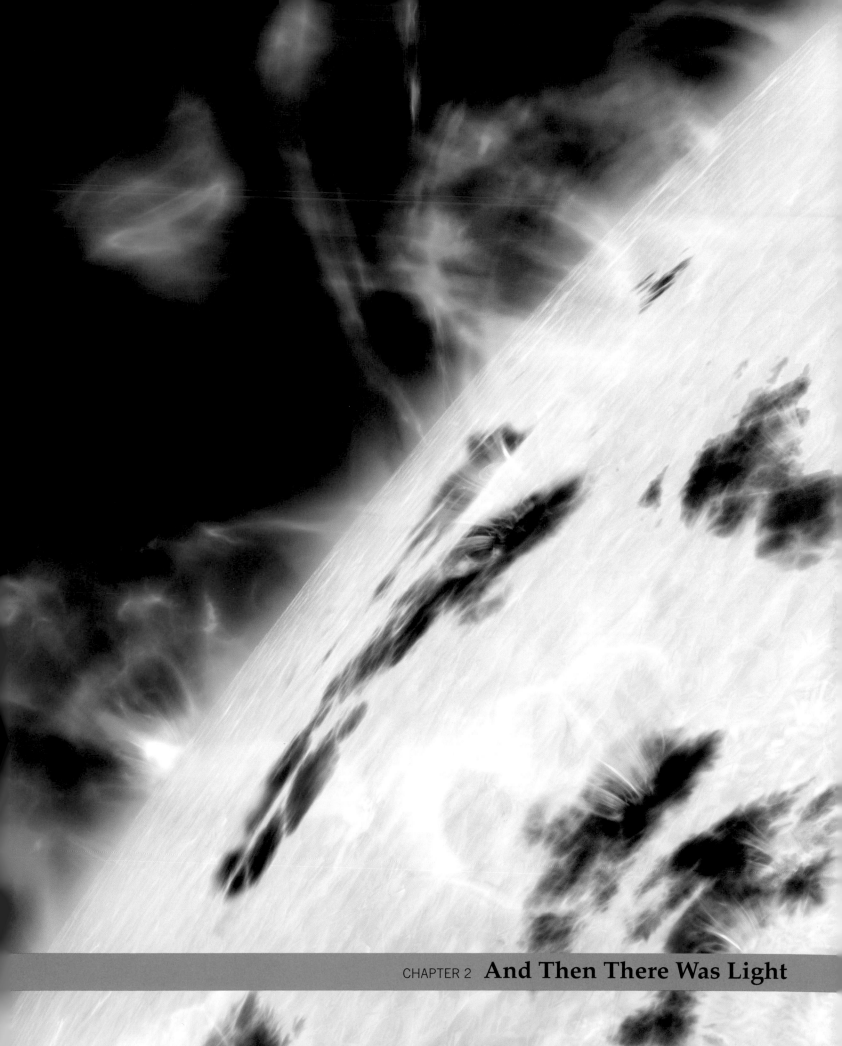

CHAPTER 2 **And Then There Was Light**

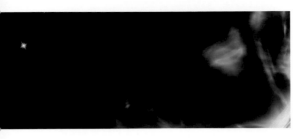

▲ The first star

The first stars are believed to have been extremely massive. Just one or two per protogalaxy would have been energetic enough to cause profound changes in their environments, paving the way for 'normal' stars like our Sun to form.

For the next 300,000 years following the cataclysmic period of inflation there were no major developments. The physical conditions that controlled the evolution of the Universe remained more or less constant. The Universe became a less violent place. As the temperature dropped, so the protons and neutrons began to slow down; however, radiation and matter were still linked, as we shall see. From our point of view, the biggest difference between this Universe and the Universe we see today is that in those very early times it was completely opaque.

Electromagnetic waves, including visible light, may also be regarded as a stream of photons, which are particles with zero mass that always move at 186,000 miles (300,000 km) per second. In the strange world of quantum mechanics (which is, perhaps, the best tested theory of modern science) we no longer have a clear distinction between 'waves' and 'particles', but have to accept that everything exists as something called a 'wave-particle duality', intermediate between the two. Just like the entities we traditionally think of as particles, such as electrons and protons, light behaves sometimes as a particle, the photon, and at other times as if it were a wave.

Each photon carries a well-defined 'quantum' of energy, the amount of energy being determined by the colour of the light, so that it is quite in order to say that electromagnetic radiation is 'a stream of photons'. Let us now follow the path of one of these photons, perhaps released by collision between a proton and an anti-proton in the very early Universe. In such crowded conditions, no photon could travel very far before hitting and being absorbed by an electron, which thus would gain energy. Eventually the photon might be re-emitted, but in almost all cases in a different direction from its original heading. This process would be repeated again and again, leaving the photons effectively getting nowhere very fast.

However, when the Universe had cooled to a mere 3000 degrees, around 300,000 years after the Big Bang, a sudden change took place. Before this critical moment, the electrons – the lightest, and therefore fastest of the constituent particles of ordinary atomic matter – had been moving much too fast to be captured by the heavier atomic nuclei, but at a temperature of 3000 degrees they could no longer avoid capture. The first neutral atoms were formed. Seen on the scale of the atom, the captured electrons orbit a long way from the nucleus (atoms are, after all, mostly empty space), but compared with the distance between atoms they are very close to their nuclei. A large expanse of space between each newly formed atom therefore opened up, and photons were suddenly free to travel for great distances. In other words, matter and radiation were separated, and 300,000 years after the Big Bang the Universe became transparent.

Echoes of the Big Bang

The capturing of the electrons was amazingly sensitive to the temperature of the Universe; as soon as this dropped below the critical value, then the process occurred with remarkable rapidity. Along with the fact that the temperature of the Universe is almost exactly the same throughout the entire extent of space (thanks, remember, to inflation) this means

▼ The electromagnetic spectrum

The visible (rainbow) part of the spectrum, the light we can actually see, is only a very small part of the overall electromagnetic spectrum. Over the last 70 years, astronomers have begun to collect information from right across the spectrum.

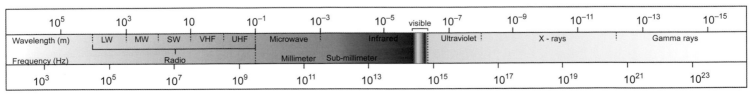

	10^5		10^3		10		10^{-1}		10^{-3}		10^{-5}	visible	10^{-7}		10^{-9}		10^{-11}		10^{-13}		10^{-15}

Wavelength (m) — LW · MW · SW · VHF · UHF — Microwave — Infrared — Ultraviolet — X - rays — Gamma rays

Radio — Millimeter — Sub-millimeter

Frequency (Hz)

10^3		10^5		10^7		10^9		10^{11}		10^{13}		10^{15}		10^{17}		10^{19}		10^{21}		10^{23}

that the process occurred almost instantaneously across the whole Universe. The result was that light could travel uninterrupted across the Universe so that over 13 billion years later we can still see a snapshot of this particular moment in the evolution of our Universe. This ability to look back to one particular instant of time is unique in astronomy. Usually, when we try to look at the distant parts of the Universe, our view is obstructed by images of nearby galaxies, that emitted their light more recently. This magical event when the universe became transparent is observable now, without obstruction, as what we call the Cosmic Microwave Background (or CMB).

Many readers will have observed these faint echoes of the death of the 'fireball' that was born with the Big Bang, whether consciously or not. By unplugging the aerial from a television or retuning it away from a channel, you will see black and white static. One per cent of this static comes from the CMB – just under 13 billion years after being emitted it is still able to interfere with your viewing of television.

When seen today, the frequency of this background radiation is consistent with an emitter at an average temperature of only 2.7 K above absolute zero. Why so cool, if this radiation is really the echo of the Big Bang itself? The reasoning is quite straightforward; the radiation would have been emitted when the Universe was at a temperature of 3000 degrees. As it travelled towards us, the space through which it was moving was continually expanding, stretching the light to longer and longer wavelengths, and hence leading to cooler and cooler apparent temperatures. This is our first encounter with the phenomenon known as redshift, which has come to be of fundamental importance.

The discovery of the Cosmic Microwave Background gave strong support to several predictions of the Big Bang theory. For instance, it was found that the radiation emitted conformed to that predicted for a black body, a hypothetical object which absorbs all the radiation that falls on it. If heated, it emits radiation in a spectrum in which the intensity of light at any particular wavelength depends only on its temperature. In practice, this

▲ The microwave sky

This all-sky picture of the sky seen at microwave frequencies reveals 13.4 billion-year-old temperature fluctuations, seen as colour differences, that correspond to the seeds that grew to become galaxies, red indicating relatively warm and blue/black relatively cool regions. The image was based on data obtained by the Wilkinson Microwave Anisotropy Probe (WMAP) satellite.

▼ Big Horn antenna

The telescope with which Robert Penzias and Arno Wilson first detected the cosmic microwave background in 1964 is more accurately described as a microwave horn antenna. It is still on show at Bell Laboratories in New Jersey (without the pigeon droppings that initially confused the astronomers).

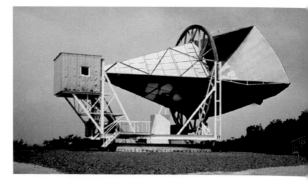

▶ The infrared sky

The top panel shows a long exposure image taken in the infrared with the Spitzer Space Telescope. At the bottom is the residual light after the subtraction of all identified foreground sources. It was recently claimed that the remaining glow contains ultraviolet light emitted by the first stars, now shifted by cosmic expansion into the infrared part of the spectrum. If the claim is confirmed, this will become one of the iconic images of astronomy.

tells us something about the nature of the emitter – for example, the object would have to be isolated from external influences. The hot, dense and practically opaque Universe of the period between the Big Bang and transparency 300,000 years later would be just such an emitter. The agreement between theory and observation is now so perfect that on most plots of the data the thickness of the line used to show the prediction is larger than the uncertainty in the measurements, a situation very rare in science and unique in observational astronomy.

At first, the radiation appeared to be absolutely uniform; there seemed to be no variations linked with direction. Even after subtracting the foreground glow of microwaves emitted by our own galaxy, one part of the sky glowing in the CMB looked much the same as any other part. But the Universe we see today is 'lumpy'; there are huge distances

▶ Redshift

We need to consider light as a wave. When this idea was first proposed there was a great deal of controversy – if light is a wave, what does it move in? After all, sound waves depend on air to be transmitted, and water waves can hardly exist on their own. Many people believed in a fundamental substance called the ether, which was all-pervading and within which all light travelled, but during the late 19th and early 20th centuries this was firmly replaced by the realization that light could be self-propagating and would have no need of a surrounding medium.

If light is a wave, therefore, it has a wavelength, which determines both its colour and its energy. Red light, for example, is of a longer wavelength and a lower energy than green light, which itself is of a longer wavelength and a lower energy than blue light. Infrared light is radiation with a

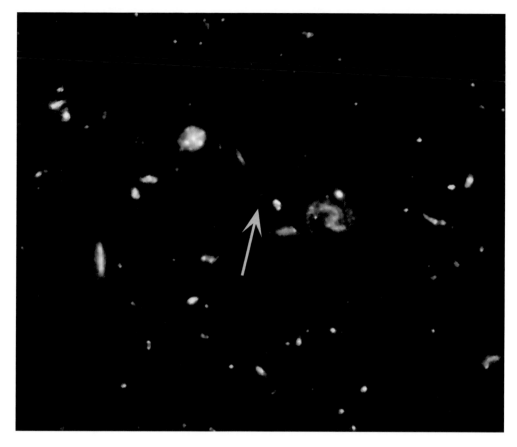

◀ **Red-shifted galaxy**

In this deep view of the Universe taken by the Hubble Space Telescope, the arrow points to a very distant, high-redshift galaxy. It appears red when compared to the relatively nearby foreground galaxies in this image.

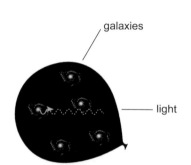

galaxies

light

1 000 000 000 years ago

500 000 000 years ago

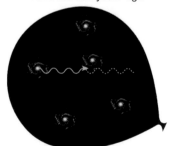

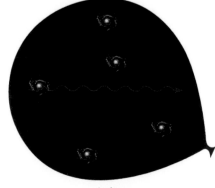

today

between the relatively dense galaxies, which are themselves grouped into clusters, and the clusters into superclusters. These superclusters are themselves separated by enormous voids, now beginning to be seen in detail in surveys such as the Anglo–Australian two degree field (2dF survey) and the Sloan Deep Sky Survey, which reach out a billion light-years from Earth. Whichever way we paint the picture of our Universe emerging from these observations, it is certainly not uniform, and so there was clearly something wrong. Hidden somewhere in the seemingly uniform early Universe there must be the seeds of the structure we see today.

The cosmic background radiation is now the most studied phenomenon of astrophysics, and it has much still to tell us. It marks our earliest view of structure in the Universe. A recent more detailed look at the CMB revealed temperature variations amounting to no

wavelength longer than that of the red light we can see, and radio waves are of still longer wavelengths. At the short-wavelength end we come to ultraviolet light, and then to X-rays and gamma rays. Since the Cosmic Microwave Background was emitted, the light we detect today has been travelling toward us through an expanding Universe. This expansion should not be thought of as objects rushing away from each other but as an expansion of space itself. As space expands, it stretches the light travelling through it, increasing its wavelength. Blue light becomes green, then red, then infrared, and we say it has been redshifted. This process can be visualized as a balloon being inflated (right). Everything on the surface moves further away from everything else. Hence the CMB, emitted in much more energetic regions of the spectrum is now detected primarily as low-energy microwaves.

▶ Traces of the Big Bang

Captured by the Cosmic Background Explorer satellite, (COBE) these maps show minute temperature differences across the sky, reflecting disconformities in the early Universe. At the top is mapped the raw data, the middle map has the effect of the Earth's motion through space removed, and at the bottom we see the result of compensating also for radiation from the Milky Way, leaving only the temperature differences resulting from the remains of the Big Bang.

more than one ten-thousandth of a degree. They may be small, but these are the ancient seeds of the structures we see around us today. It may sound strange to measure variations in density by measuring the temperature, but there is a very good reason for it. As the COBE (COsmic Background Explorer) satellite showed, matter at the time when the CMB radiation was emitted was not absolutely uniform. Regions with a density above the average, gravitationally attracted still more matter. The compression heated these regions slightly – and it is these variations that are detected and measured.

Without any fluctuations for gravity to work on, the task of producing the non-uniform, clumpy Universe we see today from a completely uniform Universe at the time of the CMB would have been impossible. However, the dimensions of the fluctuations on the sky are also important. What our observations of the CMB produce is essentially a map of the whole sky, and it is easy to see that each of the blue (colder) and red (hotter) regions appear to be roughly the same size. They are on average about one angular degree across, about twice the apparent size of the full Moon. From this single piece of evidence, and some careful thinking, cosmologists can confirm that the Universe is flat. This is possible because we know the real, physical size of the fluctuations in the early Universe; they are predicted by theory. Comparing this expected size to the apparent size tells us how the light has been bent since it left its source, and that depends on the amount of matter in the Universe. The more matter there is, the more the light is bent. In a closed Universe the light would have been significantly bent, and the net effect would have been to make the fluctuations appear larger than expected. In an open Universe – one without much matter – the fluctuations would have appeared much smaller. In fact, comparing simulation to reality reveals that the Universe has just the critical amount of matter – and is flat.

This discussion illustrates a point that is the source of both excitement and frustration to cosmologists. Excitement, because it reveals that the study of the microwave background can tell us not only about the very early moment at which it was emitted but also about the entire history of the Universe since then. This is also a problem; if we want to draw firm conclusions about the early Universe, we must be careful to disentangle more recent effects, which can be difficult to do.

The barrier of light

We have seen that before the creation of the microwave background the Universe was opaque; no light could travel far through it. We can no more look back into this era than we can look up from the Earth and see the inside of a cloud. This analogy is not perfect because a cloud is not in itself luminous; a better picture is provided by the Sun. The Sun, viewed from outside, appears to have a definite surface (the photosphere), but what we are seeing is merely the boundary at which the material becomes transparent. Inside the photosphere the gas is so hot, luminous and dense, that no photons can pass through without colliding – similar to the state of affairs immediately following the Big Bang. Outside the photosphere the gas is transparent and photons can pass through freely, similar to what happened in the Universe immediately after the event of transparency – the moment when the CMB was created.

Looking through clouds on the Earth, we have a simple remedy – radio waves easily penetrate clouds and so we can still gain some information from beyond or within them. The same trick will not work with the CMB. The 300,000 year limit applies to all electromagnetic radiation, and seems to be an insurmountable barrier. How then can we talk with confidence, as we have been doing until the last few paragraphs, about conditions before then? For now, we have to rely on our theories, many of which are able to make predictions of how the microwave background will look. We can then compare these theories with the actual CMB, and draw the appropriate conclusions.

Ideally, however, we would like to be able to look back beyond this barrier, and there are numerous proposals as to how to achieve this. We may be able to detect highly energetic particles that have survived unchanged since before the microwave background epoch. We may be detecting such particles already, in the form of tiny, almost massless neutrinos or other exotic forms of matter.

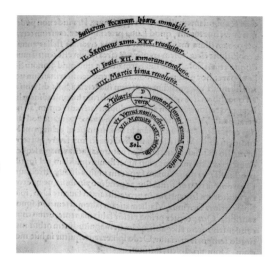

▲ Copernicus's Universe

This map by Copernicus was one of the first to show the Sun at the centre of the Solar System. A fundamental assumption underlying all of our attempts to understand the Universe is that there nothing special about our region of the Universe, and therefore we can draw conclusions about the whole from what we can see. We are guided by the so-called Copernican principle, which more formally states that no theory should place the observer in a special position. So far this has been vindicated – the Earth is not the centre of the Universe, and neither is the Sun. Neither of them is at the centre of our Galaxy, and our Galaxy is not the only one in the Universe, or even particularly distinguished.

▶▶ **How many scientists does it take to change a light bulb?**

Japan's Super-Kamiokande experiment uses many thousands of photomultiplier tubes surrounding a reservoir of ultra-pure water, to catch the flashes of light from neutrino interactions in the water. In 2001, most of the tubes exploded, necessitating large-scale repairs. Here, scientists in an inflatable dinghy check the tubes as the tank is refitted and refilled.

▼ **Ice Cube**

The latest neutrino telescope, Ice Cube, currently under construction in 2006, will employ 4200 detectors in 70 shafts sunk deep into the Antarctic ice. It is hoped that the detectors will see neutrino flashes in a cubic kilometre of clear ice. Here we see a sensor being lowered into position.

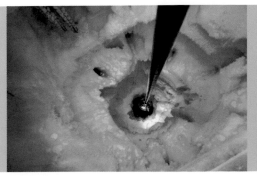

Looking back in time

Cosmologists may not be able to handle samples and submit them to analysis in the laboratory, as chemists and physicists can do, but they have one tremendous advantage: they can literally look back in time and observe the object of their study exactly as it was millions of years ago. To see further and further back in time, remember, we need only to look for objects that are further and further away from the Earth. As we have seen, this does not apply to those events before the moment of transparency, which lie hidden in the opaque infant Universe, but from now on we are discussing events that we could potentially observe directly.

This chapter began at the moment the Universe became transparent, a moment we see echoed as the cosmic microwave background. Recent experiments, such as Boomerang, Maxima and WMAP (see pages 52–53), have confirmed COBE's detection of tiny variations in the temperature of this radiation. These we interpret as an indication that there were indeed irregularities in the density of the Universe at this point in time of about one part in ten-thousand. Yet the variations in density we see around us today are much, much larger than this. We see huge galaxy superclusters, regions where thousands of galaxies crowd together and other areas of space that are almost devoid of matter.

Our own Milky Way Galaxy is just one of millions of spiral galaxies, and you might imagine that there is no reason to doubt that the galaxies (or rather, the groups of galaxies) are simply spread throughout the Universe at random. Yet large-scale surveys of galaxies reveal a wealth of honeycomb-like structure on the largest scale, including a 'Great Wall' some 30 million light-years long. How did the Universe evolve from its early, newly transparent, almost-but-not-quite uniform state to its present form?

Gravity, the universal force

The only force we would normally consider significant at astronomical distances is gravity, and the strength of the gravitational pull of an object – whether it be a star, a planet, a human being or a cloud of gas – depends on how much matter there is within it. Note that 'mass' and 'weight' are not the same thing – mass is a measure of the amount of material present, whereas weight describes the force due to gravity. Therefore an astronaut in Earth orbit is weightless but is certainly not massless. We could define gravity as being 'the force that gives mass weight'. For instance, the Moon is a relatively small member of our Sun's family, and has such a weak gravitational pull that it has not even been able to hold on to an atmosphere. The Earth is much more massive than the Moon, and therefore has a greater ability to attract objects toward it. Thus, fortunately for us, it can retain the atmosphere we breathe. Similarly, the dense regions in the early Universe had a greater gravitational pull than the regions that were less dense, and so drew in material from their

Neutrinos

These tiny particles have been studied by astronomers for the last thirty years. Quantities were produced a few minutes after the Big Bang; others are a by-product of the reactions that power stars, and are incredibly unreactive. In the course of reading this sentence, the odds are that millions of neutrinos from the Sun have passed through your body without reacting, entered the Earth and emerged from the other side of the planet. In order to study them, astronomers and particle physicists build detectors, consisting of vast tanks of liquid with which the occasional neutrino might react. These have to be built deep underground because on the surface there would be too much contamination

from particles such as cosmic rays – high-energy nuclei that slam into our upper atmosphere at close to the speed of light, propelled across the Universe by the most powerful explosions known, of which more later. In their current form, these detectors are too small to be able to function as proper telescopes. They can tell us how many neutrinos are reacting with the detector, and measure their properties, but they cannot tell us from which direction in the sky the neutrinos are coming. For that, we will need much larger facilities. One under construction is Ice Cube, which will use a cubic kilometre of the absolutely pure, transparent ice found under the South Pole as a vast neutrino detector.

▶▶ Virtual Universe

A still frame from a computer simulation of the development of the early Universe shows a slice one billion light-years across. Each filament contains the material that will clump to form thousands of galaxies, and the simulation shows that the Universe becomes more clumpy with time. This simulation includes the effect of dark matter, which interacts only via gravity. It does not, however, take into account the possible effects of 'normal' matter, which is a much more difficult computational task. Nevertheless, by comparing simulations like these with the observed reality, scientists have been able to learn a great deal about our Universe.

▼ Boomerang

The one-million-cubic-foot balloon that carried a CMB experiment into the stratosphere. It is shown here just prior to lift-off, with Antarctica's Mount Erebus in the background.

surroundings. This, of course, further increased their gravitational pull – and so on, the process accelerating all the time. Here, as has often been the case, the rich got richer and the poor got poorer!

Inside each of these denser regions there were further localized variations in density, and the same sorts of processes operated – greater mass, greater pull, more runaway collapses. Using computers, we are now able to reconstruct what went on, and to build up a model which gives a good representation of the evolution of the large-scale structure that we see in the present-day Universe.

Wherever structures are being formed, we have to consider two opposing tendencies; the expansion of space, which began with the Big Bang and, locally, contraction under the influence of gravity. Once an object in the process of formation accumulated enough mass it was able to resist the overall expansion, and collapsed.

A typical ancestor of a galaxy cluster would initially have been small, growing in volume with the expansion of the Universe, all the time accreting matter from its surroundings. As it gradually ran out of matter to accumulate, it grew ever more slowly until its expansion ceased. The embryonic cluster of galaxies had reached its maximum extent and was then able to collapse to its final size. The force of gravity weakens with increased distance, and so at this stage in the evolution of the Universe, collapse was only possible on small scales – the first galaxies, still mere agglomerations of gas, were forming.

Gloomy times

What did these aggregations look like? We can't see them as we are still looking at what Martin Rees, the 15th Astronomer Royal, has called the 'Dark Ages'. During this period, which began immediately after the epoch of the microwave background, there were not yet any stars to light the Universe.

But there was, of course, the comparatively recent echo of the moment of transparency. This radiation (which perhaps at this point we should call the Cosmic Electromagnetic Radiation Background, instead of the CMB) started life at about 3000 degrees, around the temperature of an oxyacetylene torch, so there was in fact a diffuse glow, ever fading, and becoming redder, all through this period. In fact it may be true to say the Universe was never completely dark, just gloomy!

The gravitational collapse of the material that would eventually form galaxies continued in the fading light as the Universe cooled. Then came a dramatic change; the gloom was suddenly illuminated, when multitudes of stars burst forth. The Universe exploded in a blaze of light. How sudden this was is still a matter for debate, but in any case the time has come to consider the epoch of the first stars.

Balloon with a view

Before the development of space-based research, astronomers were severely handicapped. Ground-based instruments were simply unequal to the task of measuring variations in the temperature of the microwave background. The first results with sufficient resolution to view the variations were obtained only in 1992 by the satellite named COBE (COsmic Background Explorer). New data arrived in 1999, not from space but from balloon-borne equipment carried in a helium-filled balloon and taking advantage of the dry climate of Antarctica; some astronomers believe that the southern polar region may provide the best site for astronomical observation on Earth, and further testing is under way. There were two separate projects, Boomerang (The Balloon Observations of Millimetric

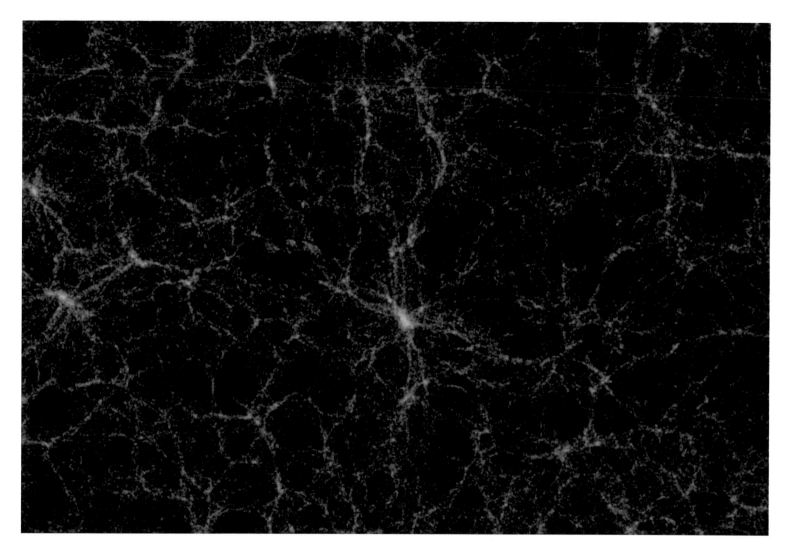

In the Big Bang itself, to all intents and purposes only three elements were created: hydrogen, helium and a smaller quantity of lithium; traces of other elements were negligible. All the other elements known to us today were synthesized inside stars. It has often been said that 'we are stardust', and this is true enough. The material in our Sun, and the entire Solar System has already been recycled through probably two previous generations of star formation. As we will see later, the explosive life history of many stars transforms the hydrogen and helium into heavier elements. The presence of gold, for

Extragalactic Radiation and Geophysics) and Maxima (Millimeter Anisotropy eXperiment IMaging Array). Boomerang had a main telescope with a primary mirror 1.2 m in diameter, and was carried by balloon up to a height of 37 km (23 miles); it covered an area of 1800 square degrees of the sky and produced a resolution some 35 times better than that of COBE. These images, which brought the microwave background into sharp focus, revealed hundreds of complex regions visible as tiny variations – each a difference of just 0.0001 degree.

Boomerang's images were surpassed by those from WMAP, the Wilkinson Microwave Anisotropy Probe. This mission revealed tiny variations in temperature; the evidence of the earliest stage of galaxy formation.

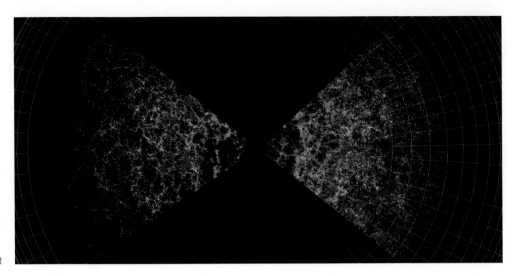

instance, is a clear indication that the material came from a supernova explosion. The first stars to form, on the other hand, were born containing only the three lightest elements.

In order to form a star, a parcel of gas must collapse, and to collapse it must cool. In the present-day Universe, radiation from carbon and oxygen atoms takes energy from the collapsing clumps of gas, but in the epoch we are describing – with no source of cooling but molecular hydrogen – the process is much less efficient. As a result, only large clumps can collapse and the stars that formed from them were also very large. The first stars were indeed extremely massive – perhaps as much as several hundred times the mass of our own Sun. With their huge reserves of fuel, one might expect these leviathans to shine for far longer than the Sun's lifetime, but in fact the opposite is true. The early stars lived fast and died young, actually surviving for just a few million years. By comparison, our Sun will have a total active lifetime of about nine billion years.

The source of stellar energy

To understand why this is, we need to consider the conditions deep in the centres of stars. Only one star is available for close study – our Sun. The Sun, like all normal stars, is a huge ball of incandescent gas, big enough to engulf well over a million globes with the volume of the Earth. Its surface is at a temperature of 5600 °C, while the core, where the energy is being produced, reaches around 15 million °C. We cannot look far into the Sun, but we can examine its constitution. We can develop mathematical models that seem to fit the observations, and so have confidence in our estimate of the core temperature. It contains a great deal of hydrogen, approximately 70 per cent of its mass. It is this hydrogen that is used as 'fuel'. And this is the same situation as in the first stars.

We have seen that a hydrogen atom, the simplest of all, has a single proton as its nucleus and one orbiting electron. Inside a star, the heat is so intense that the electron is stripped away from its nucleus, leaving the atom incomplete; the atom is said to be 'ionized'. At the star's core, where the pressure as well as the temperature is so extreme, these nuclei are moving at such enormous speed that when they collide, nuclear reactions are able to take place. Nuclei of hydrogen are combining to build up nuclei of the second lightest element, helium. Admittedly this takes place in a rather roundabout way, but in the end four hydrogen nuclei combine to make one nucleus of helium. There are also by-products;

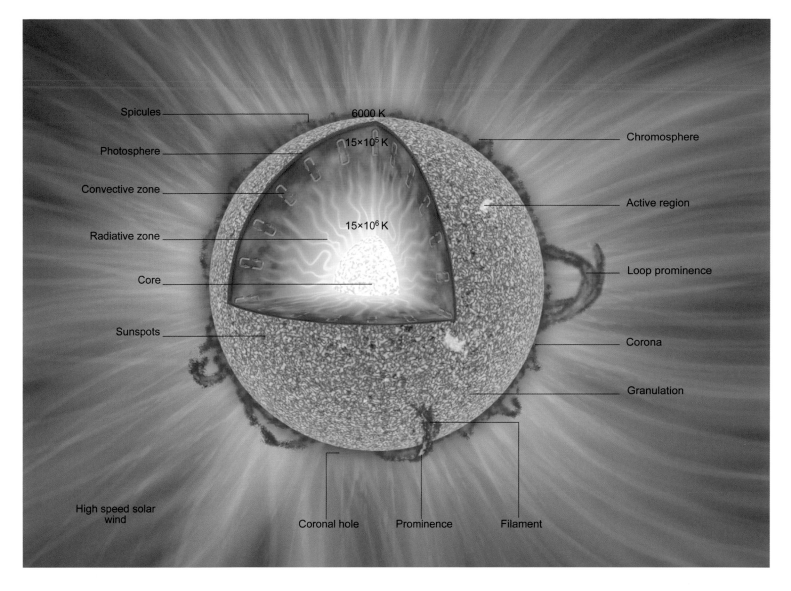

Inside the Sun

From the core to the photosphere is a distance of nearly 435,000 miles (700,000 km), approximately the same as a trip to the Moon and back.

as well as the light we receive from the stars, there are strange particles called neutrinos, of which more anon. In the process of helium-building, a little mass is lost and a lot of energy is released. It is this released energy that keeps the stars shining, and the loss in mass in our Sun is equal to four million tonnes every second. The Sun is much less massive now than it was when you started to read this paragraph. The supply of hydrogen fuel cannot last indefinitely, but there is no immediate cause for alarm. The Sun was born around five billion years ago, and as yet is no more than middle-aged by stellar standards. When all the available hydrogen has been used up, the Sun will not simply fade away; but that is another story to be told in another chapter.

So, in the Sun at least, it is the loss of mass in the conversion of four hydrogen nuclei into a lighter single helium nucleus that provides the energy that powers the star. The most famous equation in the world, $E = mc^2$, tells us that mass (m) is equivalent to energy (E). The converting factor (c^2), equal to the speed of light squared (multiplied by itself), is large, so that a tiny amount of mass loss produces a vast amount of energy. The Sun loses some four million tonnes of matter every second!

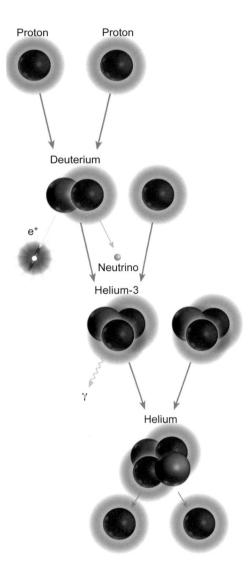

▲ Nuclear fusion

Hydrogen atoms fuse to form helium in the heart of the Sun, generating the energy that gives us light, heat and life. This process is known as the proton–proton cycle.

How does this disappearance of mass come about? Each of the four hydrogen nuclei is a single proton – hydrogen is the simplest of all atoms and consists merely of an electron orbiting a proton – whereas the helium nucleus is made up of two neutrons and two protons. However, a neutron is slightly heavier than a proton, so that if we just add up the masses of the particles on their own then it seems the helium nucleus must be heavier than four hydrogen atoms; mass seems to have been gained. Yet a helium nucleus really does weigh less than four protons, despite being composed of heavier particles. We are forced once again to remember that we are in the realm where quantum physics and its associated effects dominate, and the solution lies here. It is true that if we can measure the mass of a proton on its own it is slightly less than that of a neutron, but the subatomic particles are not free. In a helium nucleus they are bound together by the strong nuclear force and are less free to move. The creation of these bonds between subatomic particles releases energy and we measure a drop in the mass.

Why does the nucleus we produce have two protons and two neutrons? Life for astrophysicists attempting to study these reactions would be much simpler if it were possible to form a stable bond between two protons alone. This 'light helium' could then be produced by the direct, head-on collision of two protons, which would release electromagnetic radiation. However, the forces acting between two protons are not quite strong enough to hold them together when, as they both carry the same positive charge, the electromagnetic force is attempting to pull them apart. Instead of this simple picture of combining protons, the inside of the Sun, and indeed of all stars, is home to a subtle and a surprisingly slow process.

As we cannot simply add together two protons we must find a way to bypass this state, which blocks the formation of more complicated nuclei. In this discussion we only need to consider nuclei, not whole atoms, because at the temperatures that prevail at the centre of a star the electrons, which normally orbit the nuclei, making up an atom, have far too much energy to be captured. The only force that can help is the weak nuclear force, which spontaneously causes protons to decay into neutrons, releasing a positron and a neutrino. The newly-created neutron is captured by a passing proton, creating a deuterium nucleus. Deuterium is essentially heavy hydrogen, with a neutron added to the usual proton. The weak force lives up to its name, and this is the step that takes longest – a proton might spend an average of five billion years at the centre of the Sun before it is able to form a deuterium nucleus – but from here on in things speed up.

In an average of a second or so, the deuterium nucleus will snap up another proton forming a stable nucleus with two protons and a neutron – helium-3, a light form of helium. In an average of 500,000 years this nucleus will collide with another, forming the more familiar helium nucleus with two protons and two neutrons and releasing two protons, which start the cycle again. This step is delayed by the difficulty of forcing two large positively charged nuclei together. The strong force, which operates over extremely short distances, pulls the nuclei together, but they are repelled by the electromagnetic force, which keeps positively charged particles apart. Eventually the nuclei will pass close enough for the strong force to act, and we are left with energy in the form of radiation, a positron, which combines with its antiparticle and releases more energy, and a neutrino.

Neutrinos are tiny particles that move at high speed and rarely interact with other particles, and so they shoot out from the centre of the Sun relatively unimpeded by the mass of gas surrounding the core. Some of them reach Earth, where vast detectors have

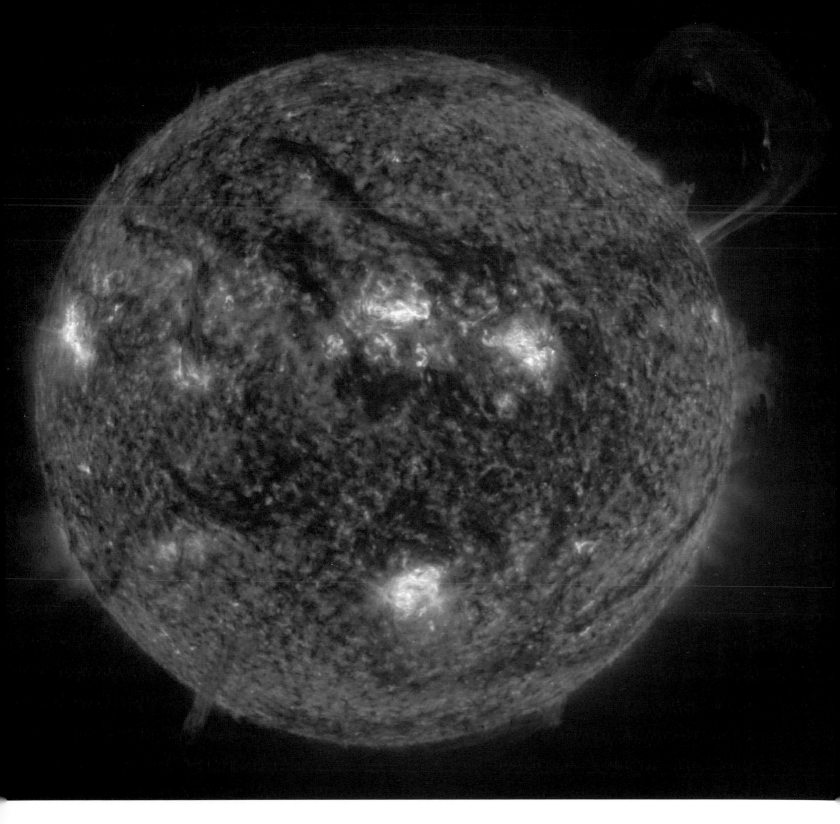

been constructed to find them. For many years, this caused a problem as there were simply too few neutrinos being detected – it was thought one must be produced for each sequence of collisions that produced a helium nucleus. However, it turns out that the neutrinos have a remarkable ability to change 'flavour', or type, en route. Particle physicists know that there are three kinds of neutrino, and it turns out they have the ability to switch between kinds over time. The original experiments were sensitive only to one particular kind of neutrino, and therefore missed all the others. These experiments confirmed that our picture of what is going on in the centre of the Sun, at far higher temperatures than any experiment on Earth could hope to reach, is basically correct. They

▲ The active Sun

This image shows a huge handle-shaped prominence – a cloud of relatively dense plasma suspended in the Sun's corona at a temperature of 60,000 degrees. The hotter areas are white, the cooler dark.

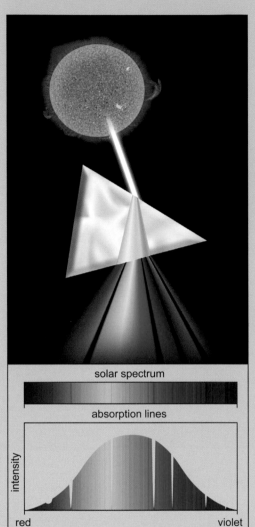

solar spectrum

absorption lines

intensity

red   violet

SPECTRA

Sir Isaac Newton was the first to pass a ray of sunlight through a glass prism, and to realize that the light was a medley of wavelengths from red (long) through to violet (short). He passed the sunlight through a hole and a prism, and drew it out into a coloured sequence – the first spectrum intentionally produced. Newton never took this experiment much further (possibly because the lenses available were of poor quality glass, but also, no doubt, because he had other things on his mind), and the next real development was due to the English scientist W.H. Wollaston, in 1801. Wollaston used a slit rather than a hole in his screen, and the spectrum of the Sun showed as a coloured band crossed by dark lines. Wollaston believed the lines simply marked the boundaries between the different colours – and thereby missed the chance of making a great discovery. The man who did so, more than ten years later, was the German optician, Joseph von Fraunhofer.

Like Wollaston, Fraunhofer produced a solar spectrum. He mapped the dark lines and found that they did not vary either in position or in intensity; for example there were two very prominent dark lines in the yellow part of the band. What caused the lines? The answer was given in 1858 by Gustav Kirchhoff and Robert Bunsen, who may be said to have laid the foundations of modern spectroscopy.

Just as a telescope collects light, so a spectroscope splits up light into its full spectrum, very much like a rainbow. Examine the spectrum of a luminous solid or liquid, and you will see a continuous band of rainbow colours. But a spectrum of a gas under low pressure will be quite different; instead of a rainbow there will be isolated bright lines – an emission spectrum (see right). Kirchhoff and Bunsen saw that each line was the trademark of one particular element or group of elements and cannot be duplicated. Thus sodium yields two bright yellow lines as well as a host of others. Some elements have complicated spectra. Iron, for instance, has thousands of lines. But the great insight was to realize that the dark lines they saw crossing the continuous spectrum of the Sun corresponded exactly to the bright emission lines emitted by glowing gases in the laboratory. We now know that each spectral line is generated by a particular transition in the state of an electron in the shell of a gas atom. If the gas is hot, we see an emission line, as the electron drops down an energy level, emitting energy, and if

▲ **Absorption spectrum**

This picture illustrates the appearance of absorption lines. The intensely hot surface (photosphere) of the Sun emits white light, which passes through the slightly cooler outer regions (the chromosphere, featuring some solar prominences is seen here). This light, being split up into its constituent colours, or frequencies, by the prism, is revealed to be made up of a continuous humped-back 'black body' spectrum, typical of an incandescent object, crossed by the dark 'Fraunhofer' lines, evidence that the gases in the Sun's cooler layers have removed these particular frequencies from the picture.

▶ **Newton's sketch**

A reproduction of an original sketch by Sir Isaac Newton, showing the layout of his famous experiment to split white light into its component colours.

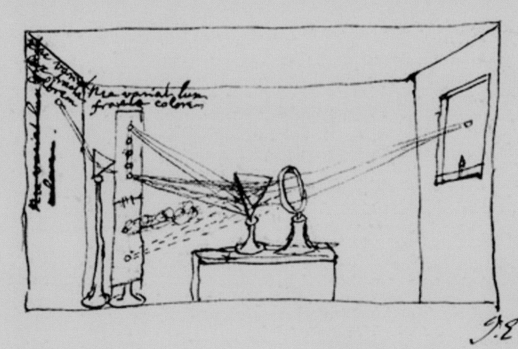

spectrum no.

1
2
3
4
5
6
7
8
9
10
11

the gas is cool, viewed against a bright continuous background like the Sun, we see a dark absorption line, since the electrons are moving up a step in energy level, and absorbing energy at this same frequency. That distinctive pair of dark lines in the yellow part of the Sun's spectrum is a clear signature of the presence of relatively cool sodium gas. From a study of these Fraunhofer lines it has been possible to establish the abundance of all gaseous elements in the Sun's inner atmosphere, a region often referred to as the 'reversing layer'.

The dark lines, now called Fraunhofer lines, can give information about motion and, indirectly, distance. Listen to an ambulance sounding its siren. When the car is approaching, more sound waves per second reach the ear than would be the case if the car was stationary; the wavelength is effectively shortened, and the note of the horn is high-pitched. When the car has passed by, and has started to recede, fewer sound waves per second reach you, the wavelength is lengthened and the note drops. This is the Doppler effect, named after the Austrian who first explained it. Exactly the same thing happens with light. For an approaching source, the shortened wavelength makes the light more blue; with a receding source the light is reddened. The colour change is too slight to be noticed, but the effect shows up in the Fraunhofer lines. If all the lines are shifted towards the red, or longer wavelengths, the source is receding. The greater the redshift, the greater the velocity of recession.

Now let us return to the solar spectrum. The Sun's bright surface, or photosphere, gives a continuous spectrum. Above it is a layer of gas at much lower pressure (the chromosphere) and this might be expected to yield an emission spectrum. In fact it does so, but seen against the rainbow background the lines are 'reversed', and look dark rather than bright. The positions and intensities are not affected; the two dark lines in the yellow part of sunlight correspond to the emission lines of sodium, and so we can prove that there is sodium in the Sun.

▲ Historic spectra

This frontispiece from Norman Lockyer's 1874 textbook *Elementary Lessons in Astronomy* (never bettered!) neatly illustrates the correspondence between emission and absorption spectra. The two distinctive yellow sodium lines are here seen on their own in emission (spectrum 5) and in absorption against a continuous spectrum (6). They are also visible below as Fraunhofer lines in the spectra of Sirius (7), our Sun (8), and Betelgeux (9). The other lines in these spectra indicate the presence of many other elements.

also provided the first firm evidence that neutrinos had a finite (although small) mass, for
if they were, as had been believed, completely massless they could not switch from one
kind of particle to another.

The life of the first stars

As the first stars to appear in the Universe – those whose light ended the Dark Ages
– were massive, each perhaps matching the weight of as many as 150 suns, the increased
gravitational pressures that came with their immense size heated their cores to very high
temperatures. The nuclear reactions that power stars must have proceeded faster, and the
material was used up rapidly. The first stars ran out of fuel in a period perhaps as short as a
million years.

Before the birth of the first stars, the Universe was a sea of atoms, mainly hydrogen. The
giant stars ignited, and their radiation spread outwards, knocking electrons out of atoms
– ionizing them. Gradually, each new star was surrounded with a bubble of ionized gas.
The more powerful stars would have produced larger bubbles. The star's energy could
only influence the gas out to a certain distance, but these stars were large enough and
energetic enough to create huge bubbles, tens of thousands of light-years across.

What happened next? Occasionally, the bubbles around two different stars met. As soon
as they did so, all the matter within them was exposed to the combined light of the two

stars. Powered by twice the energy, the bubble expanded much faster and further. This meant that there was a greater likelihood that the expanded bubble would collide with another neighbour, and the whole process accelerated. Over a relatively short period, a Universe dominated by neutral hydrogen evolved into one in which more than 99 per cent of the material was ionized.

Black holes – a one way trip

There is another possible candidate for the cause of this first ionization. (Rather illogically, this period is known as 'reionization'.) Almost every galaxy, including ours, has a massive black hole at its centre. A black hole is the product of the collapse of a massive star. It has a gravitational pull so powerful that not even light can escape from it; its escape velocity is too large. The concept of escape velocity is straightforward enough; it is the velocity an object must attain to escape from the gravitational field of a more massive body. Eventually the escape velocity of a collapsing star rises to 186,000 miles (300,000 km) per second, the velocity of light. Light can no longer break free, and since light is the fastest thing in the Universe, the old star has surrounded itself with a forbidden zone from which nothing can escape. Obviously we cannot see it, because it emits no radiation at all, but we can locate it because of its gravitational effect upon objects that we can detect – for example when the black hole is one component of a binary-star system.

▶ **Escape velocity**

Throw an object upward, and it will rise to a certain height, stop, and then fall back to the ground. Throw it faster, and it will rise higher. Throw it up at a speed of 7 miles (11 km) per second (admittedly, rather a difficult thing to do) and it will never fall back; the Earth's gravitational pull will not be strong enough, and the object will escape into space, which is why this value is known as the Earth's escape velocity. The escape velocity of the Sun, a normal star, is 386 miles (618 km) per second, while the escape velocity of the Moon, which has only an eightieth of the mass of the Earth, is a mere 1.4 miles (2.4 km) per second. This is not high enough to hold down an atmosphere; any air on the Moon has long since escaped into space. (Actually, the Moon does have an extremely thin atmosphere; it is continually replenished with dust from the surface and continually lost.) To escape from the Earth the Apollo astronauts required a massive Saturn V rocket, whereas to escape the Moon they only required the small engines on the lunar module, as seen here.

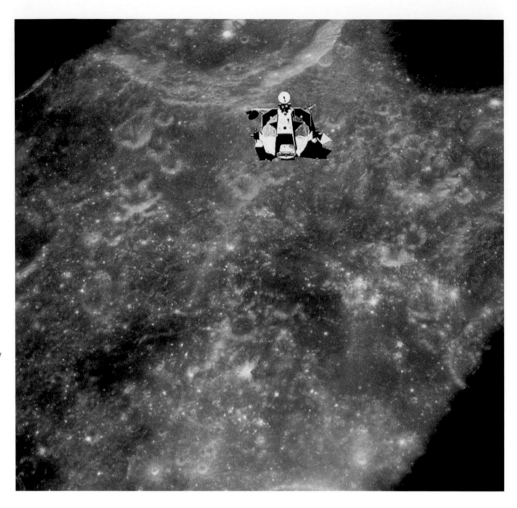

The result is that the black hole is cut off from its surroundings, and since no radiation can escape we have no way of probing the interior. We can only speculate about conditions inside. Falling into such a body would certainly be a one way trip and is emphatically not to be recommended; scientists have coined the word 'spaghettification' to describe this process – warning enough for anyone tempted to try a visit.

A black hole is usually produced by the collapse of a star, eight or more times the mass of our Sun, but this may not be true for the very large black holes in the centres of galaxies, which contain the equivalent of millions of solar masses. It may well be that these very massive black holes formed at an extremely early stage of the Universe. If this is so, then the first light may have come not from stars, but from matter heating up as it as it fell into these black holes, and this would have been sufficient to cause widespread ionization. In this case, the black holes responsible are then still with us, embedded in the centres of today's galaxies. It is not yet clear which of the two possible mechanisms of reionization is actually responsible. We need to learn a great deal more about this curious epoch before this argument can be settled.

Supernovae

Whichever theory is correct, at some point these first, curiously large stars existed, and their influence on their surroundings did not end at the time of reionization. We have

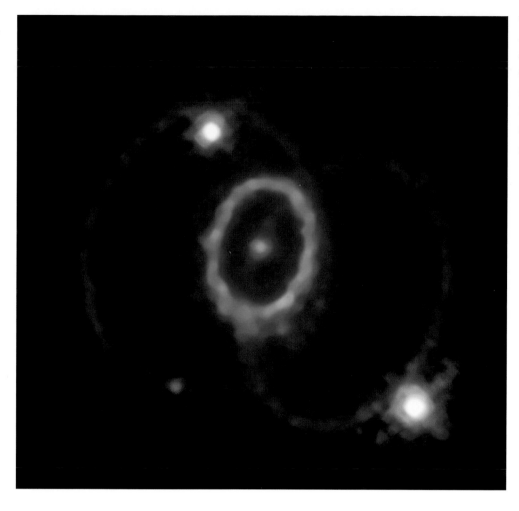

◀ Supernova rings

Astronomers are still waiting for the first observable supernova in our Galaxy since the invention of the telescope. In supernova 1987a we had the next best thing – a supernova in the neighbouring Large Magellanic Cloud. Seven years after the event, the Hubble Space Telescope imaged three extraordinary rings around the site of the explosion.

already seen that they led very brief lives; moreover, their deaths were violent. Unlike the relatively quiet future that awaits our Sun, such a massive star may be destined to suffer a cataclysmic explosion.

The outer layers of a star are supported by the energy produced in the nuclear reactions taking place in its core. When the fuel for this process is exhausted, these outer layers collapse, increasing the pressure and the temperature of the core. These changes allow helium nuclei, the product of the previous set of reactions, to collide and react with each other to build up heavier elements. Meanwhile, hydrogen around the core will still be being burnt; the result is rather like an onion with many layers, as successively heavier elements are produced in the core. Eventually iron is produced, and here the cycle stops. The nuclei of iron are the most stable of all, and therefore when they collide energy is lost rather than produced. Once a massive star forms a core of iron, nothing can prevent the outer layers from crashing inwards. A dense core quickly forms, and a shock wave rushes through the star, propelling the rest of the material outward in a vast explosion of heat and light – which we see as a supernova.

Supernova outbursts are certainly very violent. Even more extreme are hypernovae, which are created in very much the same way but involve exceptionally massive stars. Yet we have not yet witnessed the ultimate: the most catastrophic phenomena we know are called gamma-ray bursts.

▲ Supernova 1987a

Left, before the explosion on February 23, 1987, and right, 10 days after. A single supernova can outshine an entire galaxy.

Gamma-ray bursts

Gamma rays are the most energetic form of electromagnetic radiation, and have very short wavelengths – shorter even than X-rays, at wavelengths below 0.01 of a nanometre (a nanometre is one billionth of a metre). Although there is a more or less uniform background glow in gamma rays that is constant all over the sky, a few discrete sources are found. These sudden bursts of gamma rays, lasting up to a few minutes, are extremely powerful and can be seen right across the visible Universe. The initial burst of gamma rays is followed by an 'afterglow' in other regions of the spectrum, and the identification of this fading 'smoking gun' was the key that allowed astronomers to determine the distance to the more recent bursts; we now know that the gamma-ray bursts are indeed very remote.

The gamma ray story

At the height of the Cold War, military satellites were launched to look for sudden bursts of gamma rays, which are one of the signatures of nuclear testing. The American satellites sent up for this purpose did detect bursts, although they were not in the

least what had been expected. The bursts lasted for anything up to a few minutes and sometimes no more than a few seconds.

All that could be found out about them was that they seemed to be distributed evenly around the

The power emitted in a single burst is almost unimaginable – during its entire lifetime the Sun will not emit as much energy as a single burst will manage in a few minutes.

It seems that, although different bursts may be due to different causes, most gamma-ray bursts are produced by the deaths of exceptionally massive stars. Remember that once such a star has run out of fuel to power nuclear reactions, the radiation emitted from its core is switched off, and gravity finally wins the battle. The outer layers of the star rush inward, and the central regions collapse completely to form black holes. The outer layers, meanwhile, rebound and are thrown outward at tremendous speed. The energy is so great that the atomic nuclei, assembled during the star's lifetime, are ripped apart and briefly everything reverts to hydrogen. However, the energy available in this massive explosion

▲ **Crab Nebula (M1)**

This is the famous remnant of a supernova that exploded in AD 1054, observed by Chinese astronomers. Within this nebula lurks a spinning neutron star, which is all that remains of the star's core.

sky rather than being situated at one geographical location, which thankfully ruled out a nuclear test as the origin. For many years, it proved impossible to decide whether the bursts were weak, and therefore nearby, or extremely powerful and therefore extremely distant. It is now believed that these bursts emanate from sources around a billion light-years from us, and are incredibly powerful – probably the biggest 'bangs' since the Big Bang itself.

▶ **Gamma-ray bursts**

A whole-sky map of the 2704 gamma-ray bursts that were recorded over a period of nine years by the Compton Gamma-Ray Observatory. The plane of our Galaxy runs horizontally along the centre of this representation, from +180 to –180.

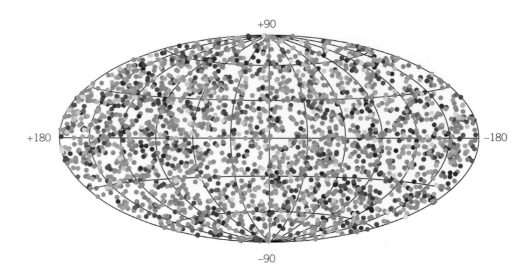

can then drive further nuclear reactions, which fuse the hydrogen atoms into heavier elements, including, significantly, those more massive than iron.

When the star involved is as large as those believed to make up the first generation, this outpouring of energy will be great enough to power a gamma-ray burst. In the nearby Universe, where the largest stars are only 20 to 30 times the size of the Sun, we see their deaths as relatively modest supernovae. The light from a single supernova, however, is still enough to outshine the entire galaxy in which it lies, and so hypernovae should be visible from right across the observable Universe.

Following this violent death, a shock wave rippled out from the explosion at close to the speed of light. A similar process can be seen in Hubble Space Telescope images of nearby supernovae. As well as heating the surrounding gas, the spreading shock wave from a dying first-generation star caused surrounding gas clouds to collapse in turn, triggering the formation of the next generation of stars. As these new stars were forming, they accumulated elements produced in the first generation of stars, which had not been available in the earlier period. These atoms, particularly carbon and oxygen, efficiently radiated away energy from the collapsing cloud. This allowed it to cool and fragment, producing smaller clumps, and eventually, smaller stars. Consequently, these second generation stars were very like those that we see today. The smallest of them (and hence those with the longest lifetimes) may even still be shining today, and we may well have detected them in our own Galaxy.

The exact mass of these stars had a profound effect on their fate. For example, stars with masses greater than about 300 solar masses would collapse directly to form massive black holes with no material expelled and no spreading shock waves. A star with a mass in a narrow band around 160 solar masses produces what is known as a pair-instability supernova. These explosions happen to produce vast numbers of positrons, the antiparticle of the electron. When particle and antiparticle meet, they annihilate, producing energy, and in these supernovae this energy is great enough to prevent the core collapsing. No black hole or neutron star is formed, and all the material is thrown outward, becoming available for the formation of the second generation of stars. We believe large numbers of stars of this size formed early in the Universe's history, and this mechanism is just what the doctor ordered.

▶▶ **Supernova remnant**

This Hubble photograph of the remains of supernova LMC N 49 shows beautiful silk-like sheets of debris that will eventually be recycled to form new stars.

▶ **Kepler's Supernova Remnant**

This combined image from X-ray, infrared and
visible light observations shows how Kepler's
supernova, which exploded in our Galaxy over 400
years ago, appears today. Among the features are a fast
moving shell from the exploded star, surrounded by an
expanding shock wave sweeping up gas and dust.

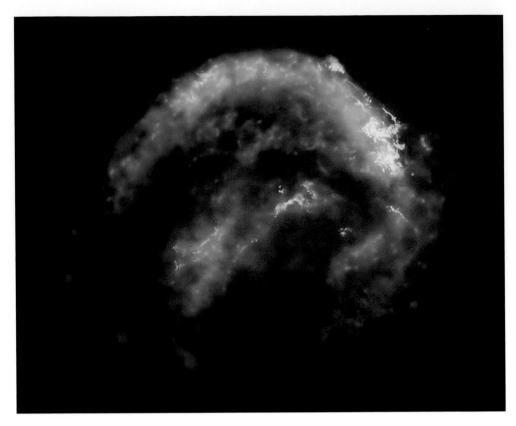

Relativity – an observer's guide

The physics of black holes is naturally written in the language of the General Theory of
Relativity, and it is worth taking the time to try to learn some of this language. According
to Einstein, if two different observers, each with their separate frame of reference, are
accelerating (or decelerating) relative to each other, their timescales will not agree. In
other words, while I may observe ten seconds elapsing, you, who are accelerating away
from me, may observe only six.

The temptation is first of all to ask who is right, and then to look for some subterfuge
that may have altered the clocks. Yet relativity tells us firmly that both are right and there
is no trick – different observers really do experience time flowing at different rates. Some
rules of common sense are preserved; two observers will always agree on the order of
events, for example. So although one may believe A preceded B by a minute, and another
that A and B were simultaneous, it is impossible for any observer to see B preceding A.
Hence cause and effect are preserved, but many other common-sense ideas that seem
second nature to us must be abandoned.

Why are such seeming paradoxes not part of our everyday experience? We never
notice clocks running at different rates, after all. The answer is that, fortunately, we don't
live anywhere near a black hole. Without extreme accelerations, or huge velocities near
to the speed of light, or very large concentrations of mass, the effects are so small that
Newton's laws of motion still work very well. Einstein did not prove Newton was wrong,
he extended Newton's ideas to be accurate in these more extreme cases.

So much for the effect of the black hole on the passage of time, but relativity also
tells us how its immense mass affects the space around it. One of the reasons relativity

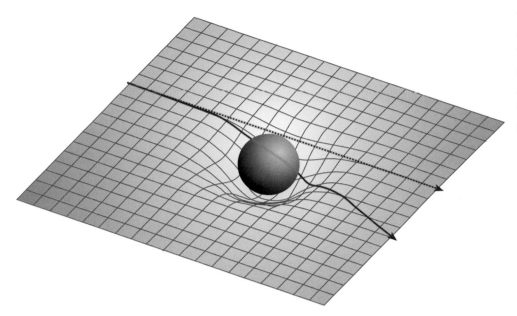

◀ **Imagining spacetime**
A massive object distorts space and time. A ray of light is bent, so what we see appears to come from somewhere other than where it actually originated. The dotted line shows the path undisturbed light would take. With a massive object in the way, the light follows the path shown by the unbroken red line.

is difficult to understand is that its mathematics is framed in a four-dimensional form – the three familiar dimensions of space plus one of time. Space and time no longer exist independently – Minkowski, who provided much of the mathematical structure of relativity, went so far as to write that 'space by itself, and time by itself, have vanished into the merest shadows, and a kind of blend of the two exists in its own right.'

Can you imagine what a four-dimensional sphere looks like? Neither can we, but we can get some idea of its properties by considering just two dimensions, picturing spacetime as a flat sheet of bed linen, being held taut at its four corners. Now, put a tennis ball or some other weight in the centre, and the sheet will be distorted, just as the theory tells us massive objects distort space and time. A light ray travelling in this distorted space time will have its path distorted and bent. Around a massive black hole this effect may be large enough to allow a suitably placed observer to see the front and back of the surrounding disk simultaneously.

Wormholes – fact or fiction?

We can do no more than speculate about conditions in a black hole. Does the luckless star crush itself completely out of existence? Some have proposed the idea that black holes distort space and time to such an extent that they could form gateways between different locations and time in the Universe, or even between different universes. This concept, known as a wormhole, currently belongs to the realm of science fiction, where it offers a useful plot device that allows characters to do things beyond the constraints of modern physics. However, it must be said that these ideas form a part of the serious academic study of these exotic objects.

Perhaps the present situation is best summarized by saying that nothing in any tested theory rules the idea of wormholes out, but there is no positive evidence in their favour, either. In any case, it seems that inside a black hole, all the ordinary rules of science break down, along with any chance of applying what we think of as our intuition based on common sense.

700 MILLION TO 9 BILLION YEARS A.B. (AFTER THE BANG)

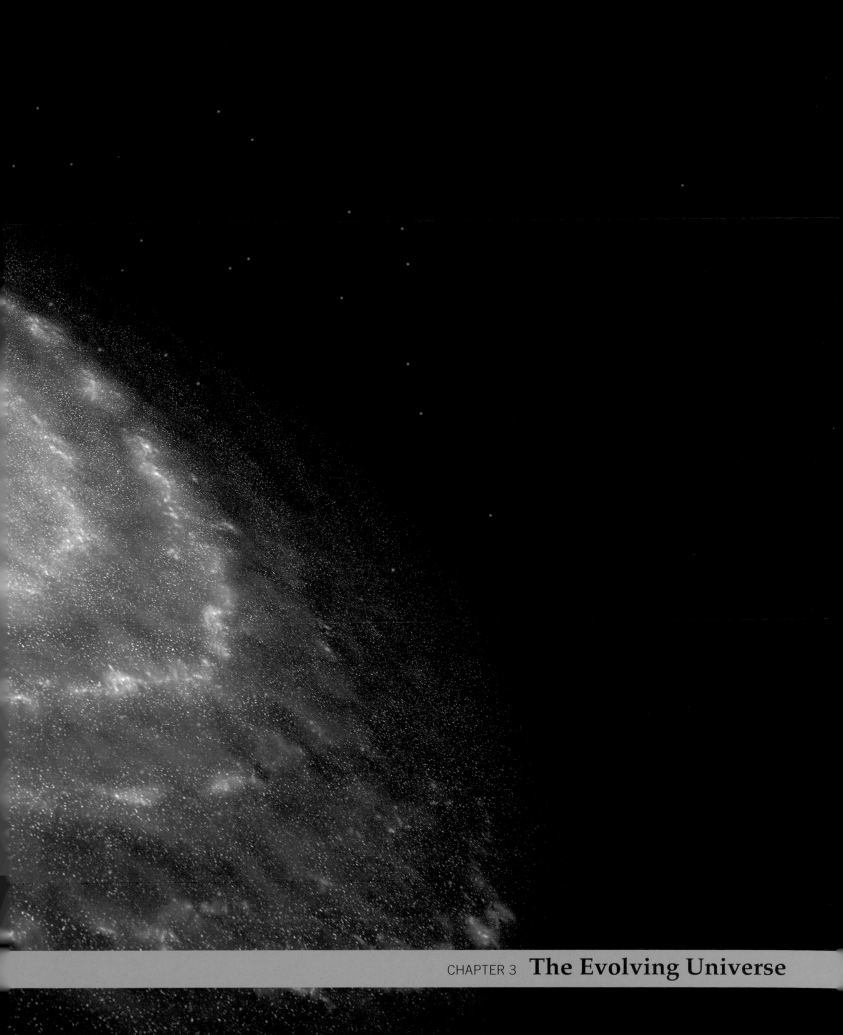

CHAPTER 3 **The Evolving Universe**

▲ The Milky Way

This artist's impression of our own Milky Way Galaxy shows the blue spiral arms, full of newly formed and newly forming stars, stretching out from the yellow bulge that surrounds the nucleus. Recent observations suggest that this bulge conceals a central bar, but the position of our Solar System within one of the spiral arms makes it hard to be sure of this.

▶▶ Galaxies unlimited

A detail of galaxies from the Hubble Ultra-Deep Field image showing only a tiny portion of the field. The complete image, itself covering only one square degree in the sky contains over 10,000 galaxies.

▶ Quasar 3C175

The term *quasar* (short for quasi-stellar object) was coined to describe a group of surprisingly energetic point-like sources of radiation that were detected in the 1960s. Such an object is now more properly identified as an extreme example of an active galactic nucleus, known to be powered by a massive black hole. Even the centre of our own Milky Way Galaxy is active to a degree, but quasars are millions of times more energetic and are only seen far away, in the Universe of the distant past. This radio telescope image shows a quasar (the central dot) emitting streams of particles (only one jet is clearly visible), which reach to over one million light-years from the centre. When the jets of particles, travelling at near the speed of light, hit the surrounding gas, they create the two blob-like shock fronts.

After two chapters, we have finally arrived at a point in the history of our Universe where there are discrete objects that we can actually see. Even before the advent of the first stars, the collapse of matter to form galaxies was in progress, and the Hubble Space Telescope's deep field images reveal galaxies just 700 million years after the Big Bang. They are not like the systems we see around us: many are smaller, and there is a wide variety of weird and wonderful shapes on view. Some harbour massive black holes. These are the mysterious quasars that dominate the scene. These powerhouses are now known to be the cores of very active galaxies shining with a luminosity equivalent to that of several thousand Milky Ways. Because they are so luminous they can be seen across vast distances, and all date back to the Universe at a fairly youthful stage.

Supermassive black holes

In the centres of these galaxies, even early on, lurked supermassive black holes amounting to several million solar masses. These could be the ones that formed directly from collapsing gas as we discussed earlier, or they may be the remnants of massive stars that have since swallowed a huge amount of extra material. In either case, a black hole this size has a gigantic gravitational pull and can attract vast amounts of material.

It seems that in the early stages of galaxy formation there was a huge amount of gas and dust available as star formation was just getting started. This material fuelled the black hole, and spiralled inward forming a disk. As it did so, it emitted light channelled into jets so that when we look down the throat of one of these jets, we see the powerful beacon we call a quasar. In this early period of the evolution of the Universe, collisions between embryo galaxies must have been common; as the two systems merge, fresh material is driven toward the black hole (or, indeed, black holes) and the quasar shines on. In fact, it may well be that all massive galaxies, including our own, went through a quasar stage in their evolution, and several quasars studied recently appear to be otherwise normal galaxies. Eventually, the fuel ran out, and the galaxies settled down.

We can glimpse this era by looking back to some of the earliest galaxies ever detected, which are shown in an image known as the Hubble Ultra-Deep Field, that was obtained by

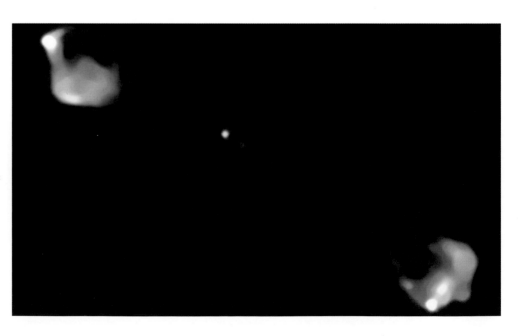

▲ Centaurus A

An elliptical galaxy 10 million light-years from Earth. This mosaic of Hubble Space Telescope images taken in blue, green and red light has been processed to present a natural colour picture. Infrared images have shown that hidden at the centre are what seem to be disks of matter spiralling into a black hole with a billion times the mass of the Sun. Centaurus A is apparently the result of a collision between two galaxies, and the left-over débris is being consumed by the black hole.

the Hubble Space Telescope. For a million seconds (just over eleven days) the orbiting observatory was pointed at a patch of sky that had previously appeared to be completely devoid of any interest. This extremely long exposure allowed the light received from even the faintest objects to build up to detectable levels, and enabled the telescope to transform the blank patch of sky into one teeming with literally thousands of galaxies. Each speck in this image represents not a background star, but a background galaxy, and although a few are relatively nearby, and look completely normal, most of them are much smaller, fainter, and frankly odder. Even by eye one can draw initial conclusions from the image; for example, galaxies that appear red are the most distant, due to their huge redshifts. Hence we can begin to put the objects detected into a rough evolutionary sequence.

From looking at these earliest galaxies and attempting this kind of analysis, we can gain an insight into the formation of the galaxies we see today. We no longer believe that each galaxy formed in isolation; if this were the case the Ultra-Deep Field would have shown a smaller number of larger, 'normal' galaxies. The new picture, initially suggested by simulations, is of early collapse leading to small structures, which merge through a series of collisions to produce larger systems. The detection of the fuel for this process in the form of the vast numbers of small galaxies in the most distant regions of the observable universe adds weight to this theory. What we may be seeing in the Ultra-Deep Field are the building blocks that make up the more familiar modern-day galaxies. This process may even still be going on; in recent years we have realized that the Milky Way is a cannibal, because astronomers have detected it ripping several dwarf galaxies apart.

These smaller systems orbit the larger galaxy, but gradually get pulled inwards. Eventually, their orbits have been distorted to the extent that they regularly pass through the disk of the larger galaxy, and on each pass their gas and dust is stripped by the larger system. After several such encounters, the smaller galaxy has completely lost its identity and becomes part of the larger system – the fate that awaits the two most obvious companions of the Milky Way, the Large and Small Magellanic Clouds.

The exotically coloured galaxies in the beautiful Ultra-Deep Field image (see pp. 6–7), which is likely to remain unique until the advent of Hubble's successor, remind us in a spectacular way of the principal evidence we possess of the central premise of this book – that our Universe is indeed expanding. The different colours of these myriad objects indicate different red shifts; the redder the object, the faster it appears to be receding from us. The light we see left them only 700 million years after the Big Bang – just five per cent of the age of the Universe. This has been verified from analysis of the position of spectral lines in these galaxies, examined in studies using ground-based telescopes.

Throughout this period, structures were still forming from matter collapsing under its own gravity, just as they had been in the Dark (or Gloomy) Ages. Among them must have been the seed that would lead to the Milky Way Galaxy, a system that is rather above the average in size, though not exceptional; its quota of 100 billion stars is exceeded by the neighbouring Andromeda Spiral. Neither is the local group of galaxies exceptional; other groups are much more populous. The Virgo Cluster, whose members are on average around 60 million light-years away, contains well over a thousand large galaxies.

▲ Hidden companion

On the far side of the central Milky Way, a small galaxy, know as the Sagittarius Dwarf Spheroidal Galaxy, has been found. It was once our nearest galactic neighbour, but is now being ripped apart by the powerful gravity of the much larger Milky Way. Peering through the Milky Way's stars, it was noticed that some of the background stars were not moving as they should. The galaxy is roughly the shape of the red region, and is a mere 80,000 light-years away. Our illustration shows a contour map of the radio wave intensities recorded for this galaxy, superimposed on a visible-light photograph of the region.

▲ Magellanic Clouds

The Large and Small Magellanic Clouds, visible only to observers in the southern hemisphere, are our second and third closest neighbouring galaxies at 179,000 and 210,000 light-years away, respectively. They orbit the galactic centre, and it seems they pass through the Milky Way's disk regularly, losing stars each time.

Our Galaxy, the Milky Way

Young galaxies contained large reservoirs of gas and dust that could be converted into stars. They are likely to have been dominated by the light from bright, young, blue stars and to look somewhat like our own Galaxy – a perfectly normal spiral. It is worth taking a slightly more detailed look at the Milky Way Galaxy before turning to others. We know it to be spiral in form, and we know that the galactic centre is about 27,000 light-years away from us. The overall diameter of the system is over 100,000 light-years, and in shape it

Discovery of spiral galaxies

One of the most important discoveries in the history of astronomy was that the misty-looking objects once called 'spiral nebulae' are other star systems well outside our Galaxy and that they are racing away from us in the general expansion of the Universe that began at the start of time and is still going on. The rate of expansion has not always been the same.

The discovery that many of the galaxies are spiral in form was made by the third Earl of Rosse, an Irish nobleman. At Birr Castle in County Offaly he constructed a reflector with a metal 72-inch mirror – much the largest ever made up to that time – and used it to examine the nebular objects; his drawing of the Whirlpool Galaxy, M51 (left) is amazingly accurate, though it was made as long ago as 1845. For many years the 72-inch reflector was out of use, but it is now in full operation again. Nothing like it had ever been built before, and nothing like it will ever be built again!

has been likened to a double-convex lens (in less scientific terms, two fried eggs clapped together back to back). Looking along the plane of the system, we see many stars in almost the same line of sight, causing the lovely band of light which traverses the night sky, known since time immemorial as the Milky Way. The diameter of the central bulge (the yolks of the fried eggs) is around 20,000 light-years. Out of the plane and away from the main disk are the vast, condensed globular clusters as well as many 'stray' stars, which inhabit what we call the 'galactic halo'.

We cannot see the galactic centre easily, because there is too much obscuring material in the way, but this is no obstacle to radio waves and X-rays. The centre lies behind the star-clouds in Sagittarius. The exact location is marked by an intense radio source known as Sagittarius A* (pronounced Sagittarius A-star). In the central region there are swirling dust-clouds, plus clusters of very powerful stars, and very near the true centre there is a black hole with a mass around 2.6 million times that of the Sun. The evidence for this comes from a very close look at 28 stars near the galactic centre. Since 1992, astronomers have watched them orbit whatever lurks at the centre, moving at orbital speeds of thousands of miles per second. By tracking the motion of these stars, we can calculate the mass and estimate the size of the central body. With so much mass packed into such a small space, the object at the centre of our galaxy cannot be anything except a black hole.

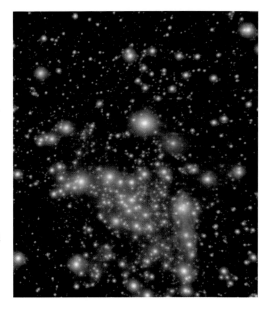

▲ **Galactic centre**

The central parts of our Galaxy, the Milky Way, as observed in the near-infrared with the NACO instrument on ESO's Very Large Telescope. By following the motions of the most central stars over more than 16 years, astronomers were able to determine the mass of the supermassive black hole that lurks there.

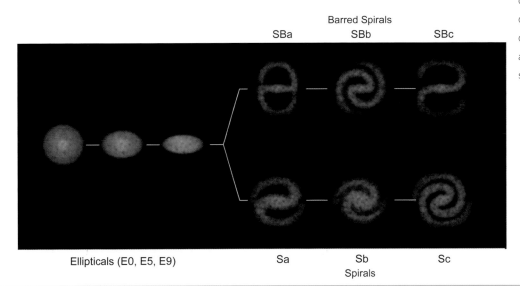

Barred Spirals
SBa   SBb   SBc

Ellipticals (E0, E5, E9)   Sa   Sb   Sc
Spirals

▲ **Classification of galaxies**

It was Edwin Hubble who drew up a system of classification for galaxies. The result is popularly called the Tuning Fork diagram, for obvious reasons – see above). Some galaxies are elliptical, while others are spirals, like Catherine wheels, and yet more are irregular.

The elliptical galaxies range from E0 (virtually spherical) through to E9 (very flattened). Spiral galaxies may be Sa (tightly wound), Sb (looser) or Sc (looser still). Some spiral galaxies show a bar through their major axes, with the spiral arms extending from the ends of the bar (SBa, SBb or SBc). Many readers may be pleased to know that it is now thought that our own Milky Way Galaxy has a bar!

It was once thought that the Tuning Fork represented some sort of evolutionary sequence, with an elliptical galaxy turning into a spiral one or vice versa. However, the picture is now known to be much less straightforward than this.

▲ Sombrero Galaxy (M104)

Named for the famed Mexican hat, this galaxy's hallmark is the dark dust lanes comprising its spiral structure. We view it almost edge-on, only six degrees north of its equatorial plane.

▶▶ Clockwise from the top: Whirlpool Galaxy (M51)

This exotic object comprises a large spiral galaxy and a smaller, barred and more amorphous companion. The perfect spiral arms of the larger galaxy may be the result of the pull of the smaller one. It was the first of the external spirals to be resolved into individual stars.

Giant elliptical galaxy (NGC 1316)

Dust lanes are visible, and some of the star-like objects nearby are globular clusters – huge star systems of 10,000 to 1 million stars. Most of the stars in elliptical galaxies such as this are old, formed more than two billion years ago.

Andromeda Galaxy (M31)

Our nearest large neighbour and the best known after the Milky Way. In this ultraviolet view, star-forming regions appear blue and existing stars yellow.

The Galaxy is rotating. The Sun takes about 225 million years to complete one orbit – a period often called the cosmic year. One cosmic year ago, the most advanced life-forms on Earth were amphibians; even the dinosaurs had yet to make their entry. (It is interesting to speculate about what our world will be like one cosmic year hence!) We travel not far from the main plane of the Galaxy, and we have just left one of the spiral arms, known as the Orion arm, so that we are now in a relatively 'clear' area.

Spiral galaxies

Many galaxies are spiral, and with one baffling exception they spin so that the arms are 'trailing'. Arms are now thought to be due to pressure-waves that sweep round the system. These are regions where the density of the interstellar material is greater than average, which triggers star formation. The stars that are most visible are very massive, and short-lived by cosmological standards before exploding as supernovae, but their brilliance makes the spiral arms dominant; when the pressure wave sweeps on, furious star formation stops, and the spiral arms become less evident. The sweeping pressure wave then creates a new arm. If this scenario is right, then in several tens of millions of years time our Galaxy will still have spiral arms, but they will be made up of different stars.

The physics that governs the spiral arms of our galaxies can be likened to a more mundane problem – traffic jams. Consider traffic on the M25, the circular motorway around London. All the cars travel at roughly the same speed, but, if the road is busy, a car travelling at a slightly reduced speed can cause a build-up of traffic behind it. This is exactly what happens to dust or gas orbiting the centre of a galaxy as it piles up in the spiral arms. Individual cars are part of the hold-up for just a limited time and will eventually move past it and on round the motorway, but the jam persists, comprised of other cars approaching from behind.

▲ Dusty spiral galaxy (NGC 4414)

Thirteen separate exposures by the Hubble Space Telescope produced this beautiful, detailed view. Older yellow and red stars populate the central regions, while the outer arms are bluer. The formation of young blue stars in these spiral arm structures is triggered by a pressure wave that travels around the disk.

▶▶ Cigar Galaxy (M82)

What is lighting up the irregular Cigar Galaxy? It was disturbed by the close pass of nearby M81, and recent evidence indicates that the red expanding gas may be being driven out by the combined particle winds of many stars creating a galactic superwind. Material caught up in this wind is moving at enormous velocities.

We have been able to measure the rotation of many spiral galaxies – mainly by the Doppler effect. If a spiral is rotating, then all the material on one side will be approaching us and all that on the other will be receding (when due allowance has been made for the overall motion of the galaxy, of course). This motion will be revealed by the position of the spectral lines and therefore the rate of rotation can be found. But there is one peculiarity that has profound significance.

Mysterious dark matter

In our Solar System, the orbital speed of a planet decreases with increasing distance from the Sun, because gravity becomes weaker further from the Sun. Logically, the same laws should apply to rotating galaxies. Stars near the centre should move much more quickly than those that are further out. Yet astronomers found, to their consternation, that this did not happen; far-out stars have a shorter cosmic year than they would be expected to, and so the spiral arms do not rapidly wind up. The galaxy seems to behave like a cross between the solar system and a solid body, acting like a spinning bicycle wheel. A speck of mud close to the hub will move more slowly than one on the rim, yet both complete a revolution in the same time.

If the stars in a galaxy simply orbited a central mass like the planets orbit the Sun, there would be no way of explaining this strange behaviour. The only possible answer is

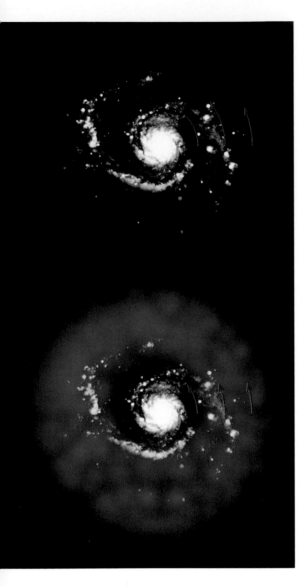

▲ Dark matter

If the matter we see is all that exists, the laws of gravity predict that the outer arms of a galaxy will rotate more slowly than those near the centre (top). In fact, observations show that all the arms rotate at the same speed (bottom). One explanation is the presence of a halo of unseen dark matter.

that the mass of the system is not concentrated at or near the centre, but must be spread throughout the disk and the outer parts of the galaxy. The most plausible explanation is that there is 'dark matter' distributed throughout the galactic halo. The dark matter is totally invisible but betrays its presence by its gravitational pull.

Can the dark matter be something ordinary, such as vast numbers of low-mass stars, so faint that we cannot see them except when they are very close, at least by cosmic standards? Certainly there are a great many stars (a very recent estimate gives the total number of stars in the observable universe as 7×10^{22}) but it does not seem that their combined mass would be nearly enough to account for the amount of dark matter.

Could the matter be locked up inside black holes? We can measure the mass of those we know about, and again the total seems to be hopelessly inadequate. (Stephen Hawking has predicted the presence of Earth-mass black holes, but they have never been discovered.) A solution that initially seemed much more promising involved neutrinos, which are fast-moving particles with no electrical charge; they are not easy to detect but are unbelievably plentiful, produced in vast quantities by the reactions that power the stars. Many thousands pass through our bodies every second. If neutrinos had even a slight amount of mass they could provide an explanation for the dark matter. We now know much more about them than we did a few years ago, and though they are not completely massless it seems certain that they cannot provide enough mass to solve our problem.

We are left with two solutions. The first of these is that dark matter may be composed of as yet unknown fundamental particles, each of a small individual mass but existing in sufficiently large quantities to explain the discrepancies. These hypothetical particles are known as WIMPs, or Weakly Interacting Massive Particles, and there are specific predictions from particle physics as to exactly what they might be. The alternative is that dark matter consists of ordinary matter, organized into massive but faint objects – such as planets, or small stars called brown dwarfs. Searches for these objects – known as MACHOs or MAssive Compact Halo Objects as they are believed to lurk in the haloes of massive galaxies – have been undertaken with few positive results. At present we await the discovery of a passing WIMP. There is worse to come when we encounter dark energy.

Is there an alternative to dark matter?

Our aim here is to give an account of a single model of how the Universe is evolving, supported by most of the currently available observational evidence. But in many ways the model is deeply unsatisfactory, relying as it does on two as yet unknown components of the Universe labelled dark matter and dark energy. Any theory which can dispose of the need for such mysterious actors on the cosmological stage deserves to be taken seriously.

The most promising alternative theories fall under the general banner of 'MoND', or Modified Newtonian Dynamics, and remove the need for dark matter by making minor changes to the theory of gravity. MoND achieved significant success in the past, accounting for the most of the evidence for dark matter on galactic scales. However, recent observations of the galaxy cluster 1E 0657-56 - the Bullet Cluster - show effects which are much harder to explain using simple MoNDian arguments.

The pink region in this image of the Bullet Cluster (top right) is the X-ray glow of hot gas, detected by the NASA satellite Chandra. The blue regions represent an estimate of where the mass in the cluster lies, ingeniously traced by its effect on the light from background galaxies. The 'normal' matter, seen in X-ray radiation, is at the centre of the

cluster, while the total mass is much more spread out. This can be explained only if the majority of the mass in the cluster is in the form of cold dark matter – the stuff that we have seen is necessary to account for galaxies having enough gravitational attraction to hold galaxies together. Here's how.

The Bullet Cluster is actually believed to be two galaxy clusters which have recently collided. Our illustration overleaf shows stills from NASA's animated movie, viewable at www.banguniverse.com, which convincingly portrays the way that this could have happened. Astonishingly, in a cluster of galaxies such as this, the gas in the vast spaces between the individual galaxies makes up roughly half the total mass. It is important to realise that in these pictures, at these wavelengths, we do not see the galaxies at all – we are 'seeing' only the gas. As these two clusters begin to pass through each other, the individual galaxies rarely, if ever, collide, because the spaces between them are so huge, but the gas molecules in the spaces do collide, scattering each other, and effectively slowing down the movement of this material. This is how 'normal' matter behaves, and the result is a large lump of normal (baryonic) matter, which we see as the central pink shape in the illustration, made up of the gas in both of the original clusters. However, if

▲ **The Bullet Cluster**

In this composite image of galaxy cluster 1E 0657-56, the individual galaxies are shown as they appear at optical wavelengths. Their total mass adds up to far less than the cluster's two clouds of hot, X-ray emitting gas shown in red. With more mass than the galaxies and gas combined, the distribution of the dark matter in the cluster is shown in blue. The dark matter was mapped by observing gravitational lensing of the background galaxies (see pages 87–9).

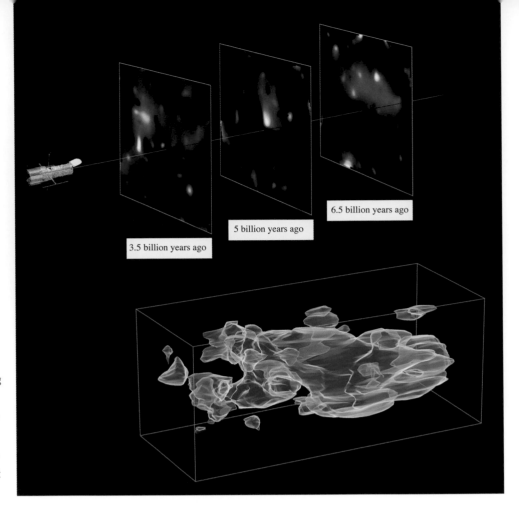

6.5 billion years ago

5 billion years ago

3.5 billion years ago

▶ **Seeing the invisible**

This 3D map is the first to show the distribution of dark matter in the Universe. It shows the filaments collapsing under gravity and growing clumpier with time. How can we see this invisible stuff? The map was constructed by looking at the shapes of half a million faraway galaxies: the dark matter deflects their light slightly, so we do not see them directly, rather through their influence on light heading towards us.

we suppose (as seems the simplest assumption) that the cold dark matter in the clusters interacts only gravitationally, there is in this case no slowing down due to collisions. Thus, as the whole process evolves, the dark matter in the clusters is able to keep moving through, and past, the agglomeration of gas. It ends up concentrated in the outskirts of the cluster, to each side of the centre, exactly what we see in the shape of the 'Bullet'.

The method used to work out the distribution of mass in the Bullet Cluster has been used for other clusters – although none have produced the dramatic results seen in the Bullet. Using the results of almost 1,000 hours of observations with the Hubble Space Telescope to take images and the Subaru telescope in Hawaii to measures distances to galaxies in the images, astronomers have even managed to create a three-dimensional image of dark matter.

Meanwhile the Bullet Cluster stands as a reminder of the fact that the concept of dark matter continues to lead to convincing explanations, and thus is still worthy of being considered a valid theory. Although arguments continue as to the exact nature of this non-baryonic material, it is now clear that even modifying gravity does not eliminate our need to believe that dark matter - matter that we will never be able to see - is out there.

Why dark energy?

According to the latest estimate, the visible Universe – that is to say everything we can see: galaxies, stars, planets – accounts for only four per cent of the energy in the Universe. Twenty-three per cent takes the form of 'dark matter' and the remaining 73 per cent is put down to what is called 'dark energy'.

Up until this point in the history of the Universe, roughly seven billion years after the Big Bang, the expansion had been slowing down under the influence of gravity. Gravity was the only force capable of making a significant difference over astronomical distances, and it is attractive, attempting to pull matter back together. We might expect the strength of the gravitational force to determine the final fate of the Universe.

The Universe was expanding in the epoch that we are discussing, and it is expanding now. But will the expansion go on for ever, or will the galaxies turn back and rush together once more in a 'Big Crunch', many billions of years hence? Everything depends upon the average density of matter in the Universe, denoted by the Greek letter Ω (omega). If Ω is greater than 1, gravity remains dominant and in the fullness of time there would be a Big Crunch. If Ω were exactly one, then the expansion would continue to slow down but would never actually stop; this is known as a 'flat' Universe. If Ω were below this critical value, the expansion would slow down but would continue for ever. As we said when we considered inflation, the evidence we have seems to suggest that our Universe is flat, but observations of supernovae of a particular kind, known as type Ia, warn us that things may be more complicated.

We can see back to this critical epoch, roughly halfway between the Big Bang and the present day, by looking for these supernovae. Why are these particular explosions so special? It turns out they all peak at the same immense luminosity and they can therefore be used as standard candles, hence allowing us to measure distances. We compare how bright the blast should be with how bright it appears in the night sky, and the difference reveals how far away the supernova is. Supernovae that appear brighter than they should must be closer than expected.

Why should all these supernovae have the same intrinsic luminosity? A supernova of this kind is believed to be due to the total destruction of the white dwarf companion of an ordinary star. The small, dense dwarf pulls so much material away from its larger companion that eventually it becomes unstable, and there is a colossal thermonuclear explosion as the dwarf blows itself to pieces. As this explosion always occurs at roughly the same critical mass, the luminosity of the blast is the same in each case. While there are some differences which depend on factors such as the composition of the explosion's 'fuel', these can be adjusted for.

▼ Very Large Array

A moveable array of 27 radio telescopes, each 25 metres in diameter, on the plains of New Mexico. The most powerful radio telescope on the planet, the Very Large Array had a starring role in the film *Contact*.

We have two ways of calculating the distances of the galaxies that hosted the supernovae; from the redshifts in their spectra, and from the peak luminosities of the supernovae – but something is wrong. The supernovae look fainter than they ought to do, and so it seems

Invisible astronomy

We know that visible light accounts for only a tiny part of the electromagnetic spectrum, and only in comparatively modern times have we been able to build equipment enabling us to study what we may call 'invisible astronomy' from radio waves at one end of the spectrum to gamma rays at the other.

Some investigations can be carried out from the Earth's surface. Most people are familiar with the huge radio telescopes such as Jodrell Bank, which are really in the nature of large aerials. One certainly cannot look through a radio telescope!

Infrared astronomy can also be carried out from the surface of the Earth. However, many of the other regions of the electromagnetic spectrum are severely blocked by layers in the atmosphere of the Earth, and this means that we have to use space-based research methods, such as probes and satellites. This is true for almost the whole of X-ray astronomy, for instance, and there is an important satellite, the Chandra X-Ray Observatory launched in 1999, which has been immensely informative in this area.

If we had to depend only upon visible light we would be in the position of a pianist who is trying to play a nocturne upon a piano which lacks everything apart from its middle octave.

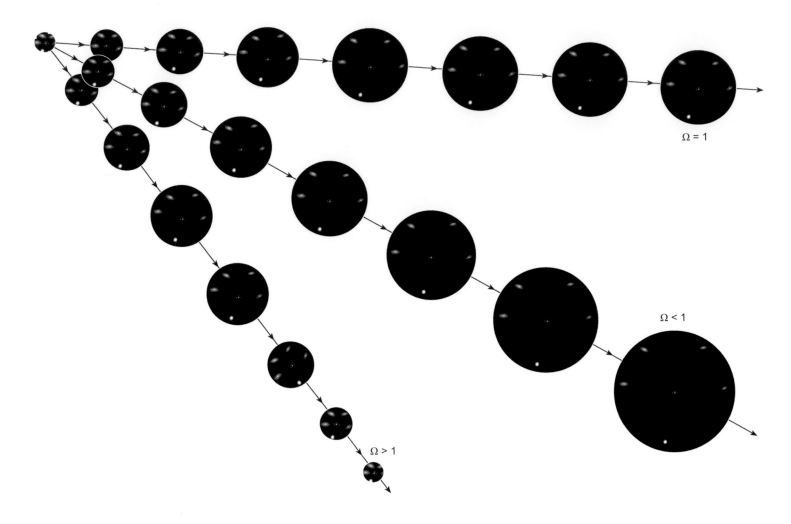

▲ The fate of the Universe

Depending on the total density of matter, the Universe will either stop expanding at some point in time and begin to shrink ($\Omega > 1$), expand forever ($\Omega < 1$), or endlessly approach – but never quite reach – a final size ($\Omega = 1$).

they are further away than anticipated. This was the last thing astronomers were expecting. Only one explanation seems possible; the rate of expansion of the Universe must be greater now than it had been before – the expansion of the Universe must be accelerating instead of slowing down. It is the energy of this acceleration that we call dark energy.

The fifth force

How can this be? Throughout the history of physics, only four forces were thought necessary in order to explain all possible interactions between matter: electromagnetic

The supernova that shouldn't have been

The first evidence for the mysterious acceleration of the Universe came from studies of type Ia supernovae, which are believed to be 'standard candles', always shining with the same luminosity no matter where in the Universe they explode. In the decade or so since that discovery, techniques have become much more sophisticated, and astronomers are now able to adjust for small

fluctuations in luminosity by studying such factors as the time taken for the supernova to reach maximum brightness and the speed of its decay. The principle has remained sound, but one recent supernova may change that. SN2006gy was decidedly odd. It was extremely luminous, and calculations showed that the mass of the material involved in the explosion must exceed the Chandrasekhar

mass of 1.4 solar masses. This is the maximum mass a white dwarf star can achieve, and so to see a type Ia supernova with more matter casts doubt on the accepted explanation for these objects. It had a very odd light curve, and would have been rejected by all of the cosmological surveys, but it is nonetheless intriguing that we cannot yet explain its extraordinary luminosity.

force (responsible for the attractive force between opposite charges), 'strong' nuclear (holding atomic nuclei together), 'weak' nuclear (causing radioactive decay) and gravity, the force of attraction which operates over the entire Universe. It is by far the weakest of the four forces, but it is dominant as far as astronomers are concerned simply because it is the only one that acts over large distances. (The electromagnetic force is also capable of long-distance interaction, but because matter is on average electrically neutral the forces cancel out.) Yet an accelerating universe required a fifth fundamental force that had not shown its effects earlier.

There are, however, theoretical speculations about a force that might fit the bill, most of which had been quickly discarded when first proposed. They lead us into the weird realm of vacuum forces and virtual particles. Quite naturally, we think of a vacuum being the complete absence of matter, but in the picture of the world we inherit from quantum physics this turns out to be an over-simplification. Any vacuum is really a seething, boiling mass of so-called 'virtual' particles, which always appear in pairs made up of a particle and an antiparticle. These virtual particles, carrying opposite charges, almost always last for only a tiny period, less than 10^{-43} seconds, before they must collide, annihilating each other. This process can be described as the vacuum 'borrowing' the energy it needs to create the particles and then giving it back by annihilating them before the rest of the Universe can notice. In their brief existence, however, these can have an effect on their surroundings – in the laboratory they have actually been seen to exert, in certain circumstances, a repulsive force. This could be just what we are looking for. Furthermore, the greater the volume of the vacuum involved, the greater the force, and so we expect the force to become greater as the Universe expands, exactly as observed.

Cosmic shear

Further evidence for the existence of dark energy has come from an unexpected source. By looking at the shapes of several hundred thousand galaxies, astronomers are able to measure the expansion of the Universe since the light was emitted from each galaxy. The method is known as 'cosmic shear' and relies on light being bent as it passes by matter. The most spectacular examples of this process are the Einstein rings, which are formed when light from a distant galaxy is so distorted by passing near a nearby system that it is spread out into a ring with the nearby system at the bullseye. Galaxies are also often seen distorted and stretched into arcs. Although these are dramatic examples, the image of every galaxy we see should be distorted in some way, and the magnitude of this distortion will reflect the amount of matter the light has had to pass to reach the

▼ **Albert Einstein**

At the blackboard in Leiden, Holland, on December 6, 1923.

Einstein's greatest blunder?

Albert Einstein believed, along with everyone else at the time, that the Universe was static, that it was unchanging on the largest scales. He realized that his theory of relativity did not allow such a universe to be stable – collapse was inevitable. He therefore introduced what became known as the 'cosmological constant' – a fudge factor that balanced gravity in the equations, and allowed the Universe to remain static. It is perhaps surprising that Einstein, whose

discoveries were based on his determination to follow his logic wherever it led him, did not believe what the equations were telling him on this occasion. Had he done so, he could have predicted the expansion of the Universe five years prior to Hubble's discovery. Once Hubble had established that the Universe was not static, the cosmological constant was largely forgotten, remembered only as what Einstein himself called his 'greatest blunder'.

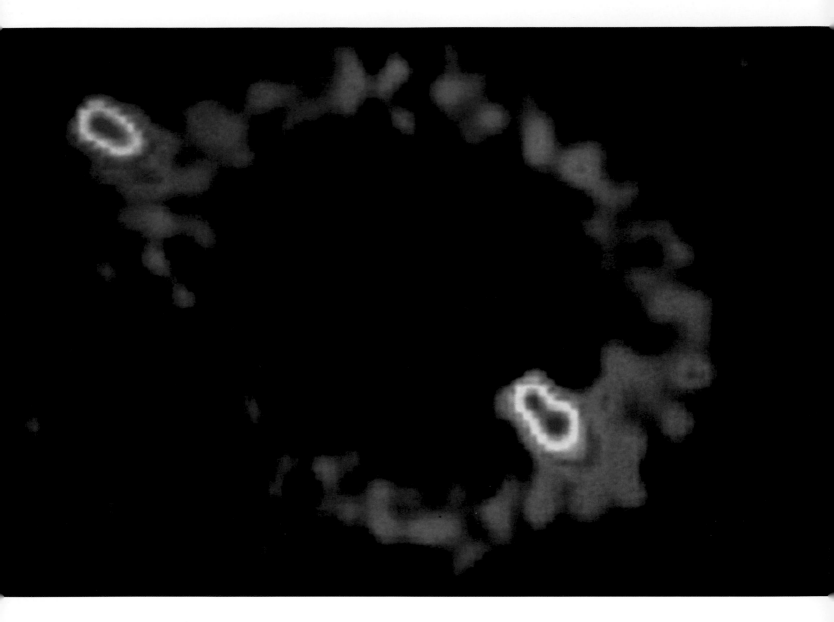

▲ Einstein ring

Emissions from the remote radio source 4C 05.51 strongly suggest that the waves have been bent by a massive foreground object, such as a quasar, to form a symmetric example of gravitational lensing. This perfect alignment, first proposed by Albert Einstein in 1936, is extremely rare and only a handful of Einstein rings are known.

observer. For most galaxies, the effect will be weak and will manifest itself only by a slight tilt in the alignment of the galaxy on the sky. There is one problem, though. We only see the galaxy after this tilt has happened, and yet to measure the amount of matter passed, and hence the expansion, we need to compare the image we see with the image as it was when the galaxy emitted the light, before any distortion. For any particular galaxy, this is impossible, but with the huge numbers of galaxies placed at the disposal of astronomers by modern surveys it is possible to take a statistical average of many galaxies and extract information in this way. The results seem to be unequivocal – an accelerated expansion is necessary to account for the path light from galaxies takes in reaching us.

There is, however, a severe sting in the tail. Before the discovery of the acceleration of the Universe, particle physicists came up with a plethora of reasons why this effect – predicted by many of their theories – did not show up in our Universe. In fact, we are left with a situation in which it seems to be possible to explain why there is no repulsive force at all, or else why there may be an extremely large effect. Unfortunately, what has

been observed is only a very small force (although added up across the whole Universe, of course, its effects are highly significant) and a major discrepancy remains. In fact, the difference between the astronomical observations and the best theoretical model is somewhere around a factor of 10^{120}, a number that has to be the largest error between theory and experiment found anywhere in science at any point in history! However, this is the best explanation we have.

The situation may be even more complicated. We have assumed that the repulsive force is constant throughout all time, and yet we have no real reason for this assumption beyond the usual desire not to complicate things. (Occam's Razor, a principle often quoted by scientists, says that when all else is equal the simplest explanation is the correct one.) Some cosmologists believe that the strength of the force responsible for the acceleration would indeed vary over time.

Many questions should be resolved in the next few decades, with further observations and new surveys already planned. It is fair to say that, for now, we remain largely in the dark!

▲ A giant gravitational lens

This amazing photograph from the Hubble Space Telescope stopped astronomers in their tracks when it was first revealed. It was an instant confirmation of Einstein's General Theory of Relativity, which had predicted that rays of light would be bent by the gravitational influence of a massive body. It is of the Abel Cluster 2218, a group of yellowish galaxies so massive that their combined mass forms a lens, producing a multitude of distorted, arc-like images of another group of galaxies much further away, seen as blueish. Sometimes multiple images of the same distant cluster are produced.

9 BILLION TO 9.2 BILLION YEARS A.B. (AFTER THE BANG)

▲ The formation of our Solar System

Material left over from the formation of the Sun makes up a disk around the young star, whose powerful jets are clearly visible. As large bodies begin to form in the disk, collisions will be common (bottom right of image).

▶▶ Giant stellar nursery (NGC 604)

In this immense cloud of dust in the nearby galaxy M33, many new stars are being born. Many of the young stars are visible not only directly, but also indirectly, as their light illuminates the rest of the nebula for us.

▶ Dark clouds into stars

This is the Bok Globule B68, photographed in visual (left), and infrared (right) light. Clouds of dust and gas are a breeding ground for stars, and this strange dark cloud, seen here in silhouette against a bright background of distant stars, could be the precursor of many protostars. As can be seen from the right-hand image, shifting to longer wavelengths allows astronomers like Chris, who study star formation, to peer through the clouds.

In the previous chapters, we saw the Universe lit up by the first stars, and we have seen the first galaxies assemble. Now nine billion years after the Big Bang, the Universe looks very much like our local, present-day neighbourhood. The galaxies are populated with second-generation stars, and it is time to say more about the evolution of stars. We already said something about the first stars, but we passed quickly over their actual formation, because we were more concerned with their all-important effects, which extended across the entire universe. We saw that they ended their short lives in blazes of glory; their supernova explosions spreading heavy elements widely. There was another effect of vital importance, too; shock waves from the outbursts triggered the formation of new stars from the neighbouring clouds of gas.

For a long time quasars were the most conspicuous objects as the black holes at their centre consumed huge amounts of the plentiful dust and gas available, liberating vast amounts of energy. As this dust and gas was used up, the quasars dimmed and the Universe was left with a population of 'normal' galaxies. Five billion years ago, the rate at which gas was converted into stars increased, and the Universe became a brighter place. Eventually though, somewhere between four and five billion years ago, the fuel began to run out and more stars were dying than were being born. Meanwhile, at just about this time in one insignificant spiral galaxy, our Sun was beginning to form, so let us therefore look at the process of star formation in rather more detail.

Birth of a star

Star formation in galaxies does not happen uniformly. The collapse can be helped by the conditions in the surrounding material, and a good example of this is in the spiral arms of galaxies such as our own. Even a brief glance at the optical image of any spiral galaxy will reveal that the stars in the spiral arms tend to be blue, whereas those in the central bulge are yellowish. The hot, massive blue stars in the arms have only brief lives on a cosmic scale, lasting for only a few tens of millions of years. This means that wherever we see blue stars, we can be sure we are looking at a region in which stars have been forming comparatively recently, and we conclude that in spiral galaxies star formation is concentrated in the spiral arms.

All stars, including our Sun, form in the immense stellar nurseries that we call nebulae, which we may regard as reservoirs of gas and dust. Inside a nebula, the material is shielded from the harsh radiation that pervades the rest of the universe, and so the matter can cool down to very low temperatures. The way in which this happens is crucial to the whole process of star formation. Initially, the cooling is due to the fact that molecules of hydrogen are able to radiate away energy. The loss of energy cools the cloud, and the temperature falls; later this job will be done much more efficiently by carbon or oxygen atoms. The collapse of these regions of gas under the influence of gravity is opposed by the random motion of the particles; if these particles are moving quickly they will overcome the grip of gravity, and the clump will never collapse enough to form a star. Modern observations of star-forming regions indicate that this may be an ongoing process, with clumps continually forming and then dissipating.

However, remember that the speeds of the particles depend upon temperature; the lower the temperature, the slower the particles will move. If the gas cools sufficiently, gravity will win the tug of war and a clump of cool gas will tend to collapse.

Once collapse has reached a certain point, it will not reverse; a protostellar core has been formed. A core of this kind will contain a great many small particles, which astronomers call 'dust'; they are about the size of sand grains and are composed mainly of compounds of carbon or silicon. It is this dust that makes the study of star formation so difficult, particularly at optical wavelengths, because visible light is almost entirely blocked by the dust. Infrared observations are useful for the view they give us of regions of warm dust. However, in the earliest stages of star formation the temperature may fall as low as 10 K, and here even infrared fails us. To observe these – the coolest places in the Universe – we must turn to the submillimetre region of the spectrum.

The temperature inside the nebula is so low that gas freezes on to dust. The gas is primarily of hydrogen, but also contains simple compounds, such as carbon monoxide. Each type of molecule forms a layer of ice. Recent work has indicated that this layered structure may be overly simplistic, and that ices are mixtures of different molecules.

▼ **Whirlpool Galaxy (M51)**

The spiral arms of this superb galaxy are the birth sites of massive and luminous stars. The image at left is seen in visible light. The submillimetre image at right shows active star formation at positions that are obscured by dust in the optical image.

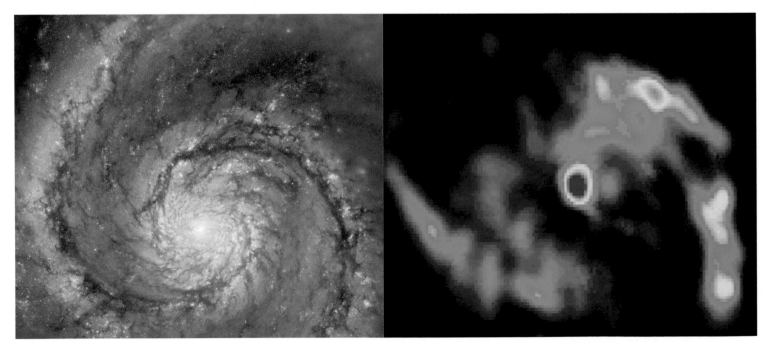

Gases at these very low temperatures move slowly; this, combined with the incredibly low densities tells us that collisions between molecules must be rare and – when they do occur – of low energy. It's worth noting that what astronomers call a 'dense' cloud corresponds to what we would normally call an extremely good vacuum in a laboratory on Earth. Therefore, relatively few chemical reactions take place.

Once the molecules have frozen on to the surface of the dust grains, the situation is quite different. Molecules are held next to each other, and there have been some suggestions that molecules or atoms (especially light atoms such as hydrogen) might move around naturally on the surface of the grains. Thus chemical reactions happen rapidly whenever molecules meet, and quite complex molecules containing ten or more atoms can be built up, all of which remain invisible to astronomers. This process is

▲ **Pillars of Creation**

From perhaps the best known image from the Hubble Space Telescope, these pillars of interstellar gas and dust are a chrysalis for new stars as well as a work of art. Several of the newly formed stars can be seen emerging at the tips of spine-like features.

important; it suggests that complicated molecules are produced as a natural consequence of the star-forming process, and they are therefore already in place when planets begin to form from the left-over debris.

Meanwhile, collapse is continuing, and the density of the central core is steadily increasing. At this stage the clump is several light-days across – a few tens of times the size of our Solar System. Eventually, the density becomes so high that hydrogen atoms collide with sufficient energy to form helium. Deep within the clump of relatively dark gas, our star has ignited. It is not yet visible, because it is still shielded by the dust around it.

As soon as this happens, the surrounding clump of dust and gas is rapidly heated, and turns into what we call a 'hot core'. This is a misnomer, far from being hot the temperature is only 300 K, about the same as Selsey (Patrick's West Sussex home in the south of England) in September. The ice is nonetheless melted, releasing the newly-formed chemicals into the gas, where they form a soup of complex molecules that can be detected by telescopes sensitive to submillimetre radiation. This stage does not last for more than around 10,000 years, which is a mere instant in astronomical terms.

The chemistry of life

Over a hundred species of molecule have so far been detected in these warm regions, many of which are familiar from life on Earth; the alcohols methanol and ethanol, for example. There are even hopes that we might have detected some basic amino acids, which make up all proteins and hence form the basis of all life known to us. If complex chemicals are naturally produced wherever stars are born, and remain in the material from which a planetary system may form, this could conceivably provide a jump start for the more complicated chemistry of life.

There is extra evidence, if only circumstantial, that the chemicals that make up life on Earth began their existence in space. So far as we know, life on Earth and on other planets is always based entirely on one atom: carbon. Each atom of carbon can form up to four stable bonds with a wide variety of other molecules, and it is this ability to bond with exactly four molecules that introduces a property known as chirality. No other molecule has quite this versatility; silicon can come close, but there is no evidence of silicon-based life appearing anywhere beyond science fiction.

Picture a carbon atom bonded to four different molecules, all of which are different. Then there are two ways of making this arrangement – each the mirror image of the other – which we call 'left-handed' and 'right-handed' forms. Both of these would have the same chemical formula and would consist of the same five components. Despite this, the two arrangements have slightly different chemical and physical properties. All simple chemical processes should produce equal numbers of left- and right-handed molecules.

How bright is a star?

Magnitude is a measure of a star's apparent brilliance. The scale works rather in the manner of a golfer's handicap with the most brilliant performers having the lowest values. Thus a star of magnitude 1 is brighter than one of magnitude 2, which is brighter than one of magnitude 3 and so on. On a clear night, an average person can see (from a dark site) down to magnitude 6 with the naked eye, while modern equipment can see down to magnitude 30.

At the other end of the scale, the brightest star, Sirius, is as at magnitude of –1.5, while the brightest planet, Venus, can surpass –4. The Sun has a magnitude of –26.7.

For complicated chemistry, such as that which takes place inside living bodies, the choice does matter. What is remarkable is that, wherever we look life seems to have made the same choice – all life on Earth uses left-handed molecules only. Why should this be so? And how did we produce a mixture of only left-handed molecules for life, when we started with an even mix? It turns out that light scattering off dust in the nebulae in which stars form may – because of a property known as circular polarization – be able to destroy right-handed particles while leaving the left-handed ones unscathed. The pattern of preferring left-handed molecules may have been set while the star was forming.

So much for the material that remains around the protostar; we will return to it later, when we consider planet formation. What of the newly-born star itself? Still cocooned in dust and gas, the violent and unstable fledgling star sends out a powerful 'stellar wind' consisting of particles thrown off from its surface, and this prevents any further material from collapsing inward. There may also be powerful jets from the poles of the stars, and these jets clear away much of the surrounding nebulosity. Around a million years after the start of collapse we have reached what is called the T Tauri stage of the star's evolution; it is still contracting, and flickers irregularly. It is surrounded by a disk of material that extends from near the new star out to several hundred astronomical units. Gradually, over the next 10 million years or so, the rest of the cloud around the star will be swept away, leaving the disk behind. The best example of this is the far-southern star, Beta Pictoris, where the disk can be easily observed once a special instrument called a coronagraph is used to block the light from the star itself.

The star in middle age

By now the star has stopped contracting, and has settled down to a stable middle age on what is known as the 'Main Sequence'. In other words, the reactions at the core provide enough energy to support the star's outer layers against the inward pull of gravity. The star is literally supported by the pressure (or 'push', if you like) of the hot gases and radiation produced in the core. Stars are so large that it takes a long time for a single

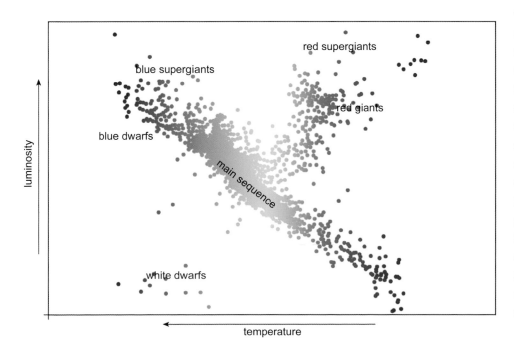

◀ Hertzsprung–Russell diagram

A Hertzsprung–Russell diagram is a way to illustrate the differences between types of stars, plotting luminosity as a function of temperature (or colour, since hot stars are blue and cool stars red, the two are equivalent). Most stars lie on what is called the 'Main Sequence', which runs from top left (hot, luminous blue stars) to bottom right (cold, faint and red). The term *sequence* should not be taken to imply that stars move along it as they age. Toward the end of their lives, most stars leave the Main Sequence, moving off to the right and becoming giants. The most massive stars leave first, then the intermediates like our Sun, which will spend a total of eight billion years on the Main Sequence, and finally the smallest dwarfs. An older cluster will lack the blue end of its main sequence, and that can be a valuable way of determining the age of clusters in distant galaxies.

▶ Forming another solar system?

The star Beta Pictoris is here seen in infrared light. It shows a surrounding bright disk of material, here seen edge-on, and with a gap in the middle due to the fact that the light from the star itself has been blocked out to make this exposure. The picture also shows an inner clear zone about the size of our solar system, evidence for the formation of planets.

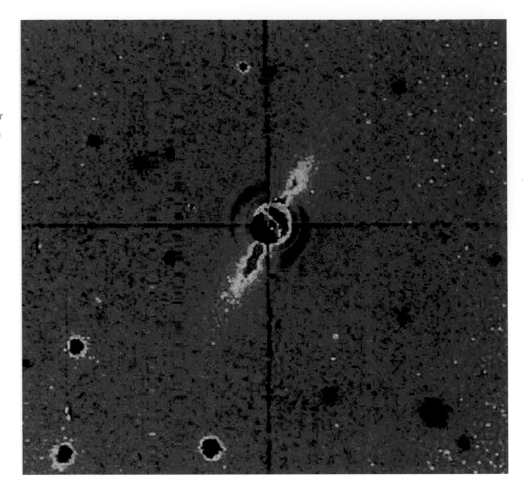

▶▶ Trapezium Cluster

A famous bright group of stars that lie at the heart of the Orion Nebula. Many 'protoplanetary disks' have been discovered in this region, primarily around what are believed to be brown dwarfs. It is thought that these discs may evolve into planetary systems.

photon – a 'particle' of light energy – to escape from the core; in the case of the Sun it may take over one million years. The whole process is regulated through a natural thermostat; if the star tries to contract under the influence of gravity the temperature in the core will rise, so that nuclear reactions occur more often, producing more energy, which forces the star to expand to its former size. Equilibrium has been reached; gravity and pressure cancel each other out, and the star can remain comfortably on the Main Sequence for thousands of millions of years.

We began by saying that stars form in immense nebulae, and then concentrated upon describing the formation of a single star. This gives a slightly misleading impression; each region of active star formation will be producing many stars at the same time, and most stars formed in these conditions will begin life as part of a cluster. A good example is the Trapezium star cluster of four brilliant young stars in the Orion Nebula, our nearest large star-forming region. Most Sun-like stars will form in binary or multiple systems, in which two or more stars form close enough together to enter into orbits around one another. Such systems can be unstable, and triple star systems often – but not always – result in the ejection, through gravitational interaction, of the least massive component. The ejection speed is usually high. A similar process occurs within clusters; stars are ejected at high velocity and, as they speed away, they carry gravitational energy with them. This loss of energy causes the remaining stars in the cluster to become bound more tightly in the gravitational pull of their neighbours, until we are left with a stable cluster. Despite these

▲ **The Solar System**

Artist's impression of our Solar System as seen looking towards the Sun from beyond Neptune. The inner planets are hardly visible even in this view, which exaggerates their size. In reality, of course, all the planets would appear as faint points of light, and the Sun as a normal star.

▶▶ **Venus**

Main image: The spacecraft Magellan mapped the surface of Venus in 1990, using radar to cut through the obscuring atmosphere. This is the Eistla Regio area and the mountain Gula Mons is a volcano standing 2.5 miles above the surface. *Below*: Because Venus is closer to the Sun than Earth is, Venus viewed from Earth appears to go through phases, like the Moon. When Venus is almost lined up between the Sun and the Earth, it is illuminated from behind, and in a small telescope, looks like a tiny crescent moon.

▼ **Mercury's southern hemisphere**

Mariner 10 was the first spacecraft to fly by Mercury, in 1974. From 440 miles above the surface, hundreds of craters are visible.

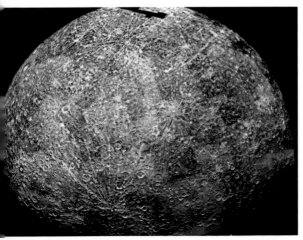

processes, however, multiple systems of some sort generally result; the single status of our Sun is one of the very few unusual things about it.

The formation of the Solar System

Around the protosun, material is left in the form of a flattened, spinning disk. The fact that the material developed into a flattened form explains why the orbital inclinations of the planets are so nearly uniform. Relative to the orbit of the Earth, the inclination is only seven degrees for Mercury and less than four degrees for the other large planets. It also explains why all the planets move along their orbits in the same sense as the Earth – if viewed from above the pole of the Sun, they would all orbit in the same direction.

Even the asteroids, and the members of the Kuiper Belt – the newly discovered swarm of small bodies in the outer reaches of the Solar System – follow most of these rules. There are no asteroids or Kuiper Belt objects orbiting in the 'wrong' direction, and of the first hundred asteroids to be discovered only four have orbital inclinations of over 20 degrees. Comets are different, as their low masses make them vulnerable to perturbations by planets, so that the orbital eccentricities and inclinations have a wide range, and comets of longer periods, including Halley's, are retrograde – that is to say they orbit in the opposite sense to that of the planets, rather like a car going the wrong way round a roundabout or traffic circle!

Researchers have developed a sophisticated model of how the disks that are observed around young stars can form into solar systems. Small, rocky planets form close to the parent star, whereas hydrogen and other light gases are driven away by the stellar wind. In our Solar System, we have Mercury, Venus, Earth and Mars and, slightly further out, the Asteroid Belt, which lies between the orbits of Mars and Jupiter. No large planet formed here because of the disruptive effect of Jupiter's gravitational pull.

Further out the situation is different. The light gases are not driven away, so once a planetary core has formed it can collect these gases and end up with a vast atmosphere; it has become a gas giant. Jupiter and Saturn are, of course, the best examples from our Solar System. What appear to be the surfaces of such giant planets are in fact the tops of their atmospheres, and this is true also of the smaller giants Uranus and Neptune.

Further out again, we come to a region occupied by much smaller bodies. Material is scarce, and the critical size needed to trap a significant atmosphere is never reached by bodies forming here. At the edge of the main Solar System we have the Kuiper Belt; Pluto is much the best-known member, though with its diameter of 1444 miles it is smaller than our Moon. The first Kuiper Belt object (other than Pluto, now generally regarded as the largest of these objects rather than a true planet) was discovered in 1992 and several hundred are now known. There are also other scattered orbiting objects, which lie further from the Sun. At least two of the objects in this gloomy region – Quaoar and Sedna – are of comparable size to Pluto.

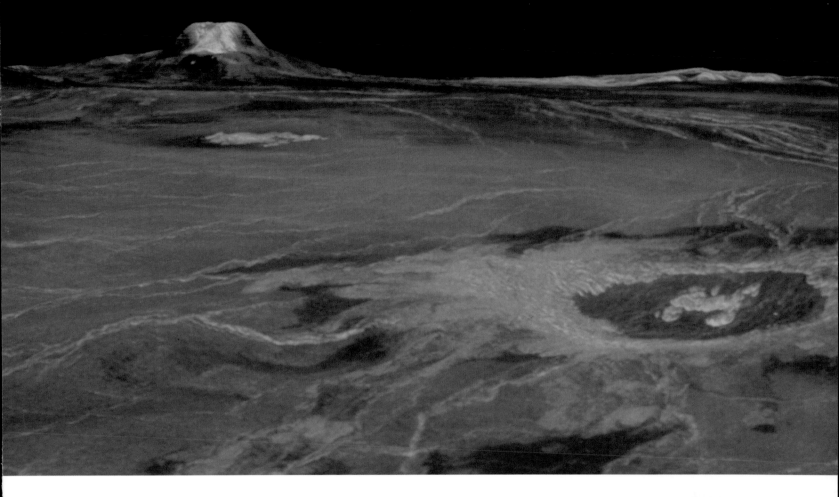

This basic picture is essentially correct but it is not the end of the story. As the gas giants form in the middle of the disk (where Jupiter lies in our Solar System) they create gaps in the disk where material has been swept up by the planet. We can observe this process in action, and gaps have been detected in the disks surrounding some young stars. In a situation such as this, a tug-of-war develops between the planet and the disk. The planet's gravity pulls material out of the disk and on to the planet, but the disk pulls back. The net result is a drag force on the planet, causing it to lose energy and thus spiral inward toward the central star.

Once a giant planet begins migrating inward it can be hard to stop, and it is a major challenge to develop a theory that can move these planets to the inner regions of the forming solar system, but prevent them suffering a fiery end as they fall into the star itself. There have been suggestions that in some cases this is exactly what has happened, and we are detecting only the latest in a long line of planets to have formed and moved inwards to destruction. More promisingly, recent work has shown that giant planets may eventually win their battle with the material in the disk, capturing all of the material nearby and preventing further drag. At this point, the migration stops, and the giant planet has found its permanent home.

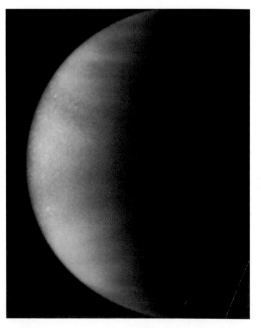

▶▶ Martian ice

Ice within a crater near the Martian north pole as seen by the stereo camera on the European Mars Express spacecraft. The Martian surface may have vast reservoirs of water ice stored beneath its deserts, providing potential homes for simple life forms.

Our Solar System appears to have escaped the chaos of a large planet ploughing through the inner disk, but this is not to say that everything was stable from the beginning. It may be that successive Jupiter-sized planets formed and migrated inwards, only to plunge into the Sun and be destroyed. Whether these planets existed or not, eventually two large clumps of material would have formed. Big enough to capture hydrogen gas with their gravitational pull, these two clumps were able to quickly increase in mass, forming Jupiter and Saturn. (Saturn's most striking feature, its rings, can only have been formed in the last million years or so, perhaps by the break-up of a moon after a spectacular collision. Inherently unstable, they will only last for another million years; we are truly lucky to be able to enjoy them.)

Near proto-Saturn, meanwhile, two more clumps were condensing from the disk. Much smaller, these clumps were able to capture some gas, but only at a greatly reduced rate; Uranus and Neptune, the planets that would form from these clumps, are just above the critical mass that divides rocky planets from gas giants. At first these planets were much closer to the Sun than they are today, but the gravitational influence of Jupiter combined with interactions with the disk caused them to swing outwards towards their present positions. This had dramatic effects. Much of the material left over in the outer disk, too cold and not dense enough to form planet-sized clumps, came too close to either Uranus or Neptune and was thrown out of its stable orbit. Much of it ended up in the furthest reaches of the Solar System, in what we now call the Oort Cloud – a vast reservoir of material located a substantial fraction of the distance to the nearest star, safe from the disruptive gravitational influence of the planets.

Occasionally, the material in the Oort Cloud is disturbed, either by interactions between the Oort-Cloud bodies themselves or by a passing star, and material is thrown into the inner Solar System, where these refugees appear to us as comets, shedding their icy material under the glare of the Sun. Such events are rare now, but in the epoch we have reached in our story, they were much more common as Uranus and Neptune flung material inwards. We see the traces of this 'great bombardment' in the record of craters on

▼ Looking down on Earth

Clouds and sunlight over the Indian Ocean, as seen from the Space Shuttle Discovery in 1999.

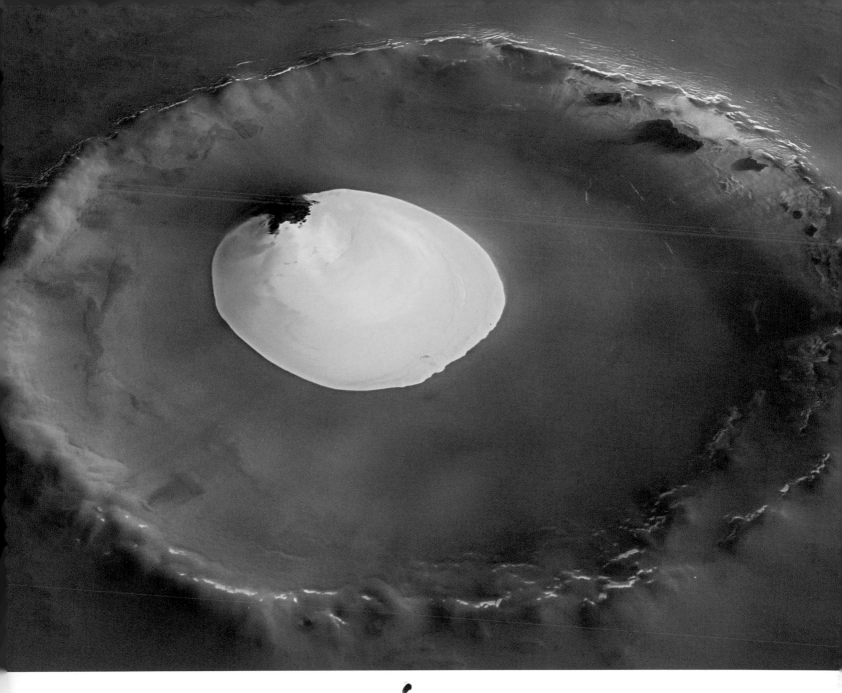

the Moon, which show that the inner Solar System was hit with a vast number of small bodies. They must have hit the Earth, too, but the traces have long since been covered up.

The Solar System today

It is most unlikely that our Solar System is unique, but it may be rather unusual, so let us take a more detailed look at it. As well as the planets and asteroid-sized bodies, we have the comets, which have been described as dirty ice balls. The only fairly substantial part of a comet is its nucleus, made up of ices mixed with rubble. When a comet nears the Sun, the ices evaporate and the comet develops a head, and very often a long tail or tails. There are also dusty particles – really cometary debris, which produce meteors when they enter the upper part of the Earth's atmosphere and burn away at heights of around 40 miles (65 km) above sea level.

Larger bodies may land intact and produce craters; these are meteorites. It should be noted that meteorites are not simply large meteors; the two classes of object are quite different. Meteorites are bodies that have been dislodged from the Asteroid Belt, and have no direct connection with comets.

▼ **Mars**

The Hubble Space Telescope's view of the Martian disk. As well as the ice caps, clouds are visible. The dark areas, once thought to be vegetation, are now known to be regions where dust has been blown away.

▲ Jupiter

Jupiter's Great Red Spot has been visible since Galileo first observed the planet. The enormous storm, larger than the Earth, was captured here by the Cassini spacecraft on its way to Saturn.

▼ Saturn

The most beautiful of all the planets, mostly due to its rings. This image, taken with Patrick's own telescope, also shows numerous bands on its gaseous surface.

The planets move round the Sun in orbits that are not far from circular; most comets, however, have very eccentric orbits. The planets have orbital periods ranging from 68 days for Mercury to nearly 165 years for Neptune. As we have seen, the planets formed from a flattened disk of material surrounding the youthful Sun, which is why their orbital inclinations are much the same. This is also true for the Kuiper-Belt objects and the comets.

The most famous comet is, of course, Halley's, which will be back once more in 2061. At the moment it is too faint to be seen, but no doubt it will be picked up long before the next perihelion (position nearest to the Sun). The really brilliant comets seen occasionally have much longer periods; some, although none so far in our lifetimes, have apparently become bright enough to cast shadows. Finally, the Solar System contains a vast amount of interplanetary dust.

Of the four inner planets, Earth and Venus are much the same size; though they are near twins in size and mass, they are nonidentical twins. Venus has a very dense atmosphere made up chiefly of carbon dioxide, and its clouds are rich in sulphuric acid. The surface is at a temperature of almost 500°C. Earth-type life there seems to be wildly unlikely. The innermost planet, Mercury, is too small to retain an appreciable atmosphere. Beyond the orbit of the Earth, we come to Mars. Many spacecraft have been sent there, and there are already plans for sending crewed vehicles, though this lies in the far future.

It is very clear that the giant planets differ totally from the small inner planets. They were formed at a greater distance from the Sun and were therefore able to retain light gases, notably hydrogen. Jupiter and Saturn certainly have silicate cores at high temperature surrounded by layers of liquid hydrogen above which comes the atmosphere we can see. Uranus and Neptune are not the same – they are better described as ice giants rather than gas giants. Jupiter is more massive than all the other planets combined,

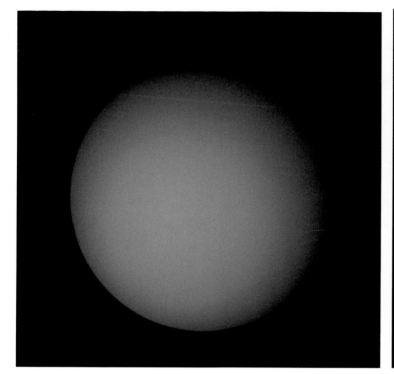

and it has been said that the Solar System consists of nothing but the Sun, Jupiter, and assorted debris!

Of the satellites of the planets, our Moon is unique in being the only large satellite to be associated with a small planet. Of the rest, Jupiter has four large satellites and a host of smaller ones; Saturn has one large attendant, Titan, and several of medium size as well as the dwarfs; Uranus has five satellites of fair size, and Neptune has one, Triton, as well as a swarm of other small satellites. Of all these satellites, only Titan has a dense atmosphere. Mars has two very small moons, Phobos and Deimos, which are certainly ex-asteroids captured long ago. Of the planets only Mercury and Venus are solitary travellers in space.

▼ **Uranus**

This rather boring surface belongs to Uranus, shown here as Voyager 2 flew past. Uniquely among the planets, the axis upon which Uranus rotates is tilted by just over 90 degrees, so it rolls around the Sun.

▲ **Neptune**

The most distinctive feature seen by Voyager on Neptune was the 'Great Dark Spot'.

The status of Pluto

In 1931 Clyde Tombaugh, at the Lowell Observatory at Flagstaff, discovered a new member of the Solar System, orbiting at a greater distance than Neptune. It was named Pluto, and was thought to be comparable in size with the Earth; it had an unusually eccentric and inclined orbit, with a period of 248 years. At perihelion it was closer-in than Neptune, but its orbital inclination (17 degrees) ruled out any danger of collision. It was accepted as a planet, but doubts about its status soon arose, particularly when it was found to be much smaller than estimated, and to be smaller than the Moon – its

diameter is now known to be 1444 miles. Moreover it is accompanied by a companion, Charon, whose diameter is half that of Pluto itself and whose orbital period is the same as Pluto's rotation period (6.3 days). Altogether, Pluto was regarded as an oddity.

Then, in 1992, a second, smaller body was found orbiting beyond Neptune; its catalogue number was 1992 QB1, and for some reason or other it has never been given an official number – it is popularly called 'Cubewano'. Other trans-Neptunians followed, some of them comparable in size with Pluto and one, Eris, definitely larger. Many years earlier

Gerard Kuiper had predicted the existence of a trans-Neptunian swarm of asteroid-sized bodies, and the swarm makes up what is now called the Kuiper Belt. In it, Pluto is by no means exceptional. Clearly, it could no longer be classed as a planet, so it was given an asteroidal number (134340) and demoted. Arguments about this took place. Finally, in 2006 the International Astronomical Union decreed that our largest Kuiper Belt objects (Eris, Pluto, Makemake and Haumea), plus the largest Main Belt asteroid, Ceres, should be classified as 'dwarf planets', the rest as 'Small Solar System Bodies'.

▲ Comet McNaught

Seen from the European Southern Observatory's
Paranal Observatory in Chile. The fan-like appearance
of the comet's tail is reminiscent of an aurora.

▼ Comet Tempel 1

This high-resolution image was assembled by Brian
from images returned by the Deep Impact probe in July
2005. The probe deliberately crashed into the surface.

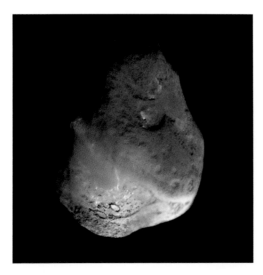

Titan

Titan is the only planetary satellite to have a substantial atmsphere, made up mainly
of nitrogen (over 98%), with some methane. It was surveyed by the Cassini spacecraft;
Cassini carried the Huygens lander, which was separated on December 25, 2004 and
made a successful controlled landing in the region now called Xanadu, a flattish plain
with 'pebbles' and dunes; the surface had the consistency of wet clay. The dunes must be
organic, and there are channels where liquid methane flows. The data from Huygens were
relayed by Cassini, which orbited the planet, though of course contact was lost when the
lander went out of range. Cassini then went on to make a long-term study of Saturn and
the satellites, including Titan.

The surface temperature is about –180°C; a methane drizzle may be almost contant.
There are hills and valleys, but few craters. Cassini has sent back images of what are
undoubtedly chemical lakes (methane and ethane); one of these, in the south polar area,
has an area of 20,000.square kilometres. It is known as Lake Ontario, and is slightly larger
than its Canadian namesake.

Life on Titan cannot be entirely ruled out, but the low temperatures and the
environment in general seem to make it very unlikely.

Enceladus

Enceladus is one of Saturn's smaller moons, not much larger in diameter than the width
of the British Isles, but it has proved to be a fascinating world. Its icy surface contains
areas which are relatively crater free, and hence likely to be young. In fact, Enceladus
is only the fourth known active world (along with the Earth, Io and Triton); a plume

of water has been seen above the satellite's south pole region. The water presumably comes from an internal ocean or lake, but the presence of any liquid within such a small body is very surprising indeed. Whatever the cause, it appears that Enceladus has an important rôle to play in the Saturnian system, as material from the moon appears to be fuelling some of the planet's more tenuous rings.

Rocky planets

If the inward migration of gas giants is common, it may be that our chances of detecting rocky, Earth-like planets are significantly reduced. Even if they formed early in the history of a solar system, they are likely to have been thrown out of their orbits or destroyed by the passage of a Jupiter-sized planet through their neighbourhood. The Earth's existence seems to depend on the fact that, for reasons not well understood, Jupiter stayed where it originally formed. Indeed, at the time of writing the vast majority of the detected solar systems seem to have gas giants where we would have expected rocky 'Earths'. It must be admitted that our techniques are biased toward the detection of large planets close to their stars, and further observations may show that our Solar System is not after all so unusual. This is a fundamental question, which we should be able to answer within the next ten years; missions are on the drawing board to look for other Earths directly.

Just occasionally observers can be lucky, and as seen from Earth a planet will appear to pass in transit across the face of the star around which it is moving. In our own Solar System we see transits of Mercury and, more rarely, Venus. The last transit of Venus occurred in 2004; the next will be in 2012, then there will be a gap of more than a century before the next pair of transits. The extrasolar planets (or exoplanets) orbit stars that are too far away to show disks, so when the planet passes in front of the star all that we see is a slight dimming as the planet blocks some of the star's light. The transit method lends itself to large-scale surveys; tens of thousands of stars can be monitored in a single night, and any suspicious slight dips in brightness can be followed up. This type of astronomy is no longer the sole domain of the professionals – it is an amazing thought that these signs of planets around other stars are now detectable by amateur astronomers. Indeed, amateurs have been credited with the co-discovery of a handful of the known exoplanets.

Extrasolar planets

We now know of more than 300 exoplanets; that is to say planets orbiting stars other than our Sun. All but a handful have been discovered by a variety of indirect methods such as the transits described above. The most successful method, though, involves looking not at the planet, but at its parent star. Although, as in our Solar System – where the Sun contains much more than 99 per cent of the mass – the central star is much more massive than the planet, the gravitational pull of the planet on the star will have an effect, causing the star to wobble as it moves through space. The wobble will be very small indeed, but careful measurements can detect it. By this method, the presence of the planet can be determined, and an estimate of its mass obtained. The greater the planet's mass, the greater the wobble, and so as with the transit method it must be admitted that there is a bias toward large planets close to their parent stars.

We now know that extra-solar planets are common. Obviously we cannot see them as anything but specks of light, but the first detection of the atmosphere of a 'hot

▲ The fountains of Enceladus

Backlit by the Sun, these fountain-like sprays of what is believed to be liquid water tower above the south polar region of Enceladus. Cassini made this discovery.

▼ Observing the transit of Venus

Brian is pictured in Patrick's garden using his own 'high-tech' equipment to observe the transit of Venus across the face of the Sun in 2004. Venus is the small dark dot near the bottom, about to leave the Sun's face. Note that we are seeing only an indirect image of the Sun, projected by the eyepiece of a small telescope (not seen) on to a piece of white card (supported by two wire coat hangers!) **It is highly dangerous to attempt to view the Sun directly through any kind of telescope. Even a small finder scope can ruin your eyesight forever.** See page 160 for full information on how to observe the Sun safely.

Jupiter' has been made. The best studied example, HD209458b, is a transiting exoplanet, and so the spectrum of the star without the planet can be obtained while the planet is behind the star, and this spectrum can then be removed to leave the spectrum of the planet itself. The result is a picture of a planet which is not like any world in our Solar System. The intense ultraviolet radiation from the star heats the planet's atmosphere, causing it to inflate. In fact, the atmosphere is escaping from the planet at a rate of 10,000 tons per second (about three times the flow rate of the Niagara Falls). In the last few years, it has become possible to study the composition of some of these hot Jupiters. For example, the closest transiting planet to Earth, HD189733b, 60 light years away, shows the signature of carbon dioxide in its atmosphere, just as our Jupiter does.

These indirect studies are of immense importance, but it has to be admitted that it would be more satisfying actually to see the planets. To make an actual observation of an extrasolar planet is obviously difficult, because the planet will shine only by reflected light and will be lost in the glare of its parent star. Two teams of astronomers managed to do exactly that, releasing their results on the same day in November 2008. Both teams used an instrument called a coronograph to block the light from the planet's parent star, revealing a system of three planets around the young, massive star HR8799 and a planet-sized object moving in the disk around the bright star Fomalhaut, 25 light years away. Fomalhaut, in Piscis Australis (the Southern Fish) is a white star 20 times as luminous as the Sun; the planet (Fomalhaut b) is more like Jupiter than like the Earth. Its distance from the star is 115 times as great as the distance between the Earth and the Sun. These remarkable images will do down in history as our first glimpse of worlds beyond our own solar system.

Such studies are necessarily in the early stages, and it is difficult to draw reliable conclusions from studies of just a few planets. In the next decade or so the first Earth-sized planets should be detected, and it is not too much to hope that we will also be able to measure the composition of their atmospheres. If so, we may be able to see the presence of an enhanced abundance of oxygen, believed to be a signature of life.

Dim brown dwarfs

There is an essential difference between a planet and even the coolest brown dwarf star. A true star must have a mass at least 8 per cent of that of the Sun, which is roughly 75 times the mass of Jupiter; below that value, nuclear reactions cannot be triggered off, because the core temperature is not high enough. Because brown dwarfs are so dim they are not easy to find, and it was not until 1995 that the first positive identification was made, but by now many have been located. Most are associated with normal stars, possibly because these are easier to locate than isolated dwarfs. To date, the dimmest known brown dwarf is known as Gliese 570D, 19 light-years away; its surface temperature is a mere 753 K, slightly greater than that of a domestic oven. It is in orbit round a triple star system, and is thought to have a diameter of about the same size as Jupiter, but it is 50 times as massive and is too heavy to be classed as a planet. On the other hand, neither can it be classed as a bona-fide star, because its atmosphere has been found to show traces of lithium – and lithium cannot survive at the temperatures of ordinary stars: it is broken up. At least the dwarf shines feebly, whereas a planet depends on light reflected from its parent star.

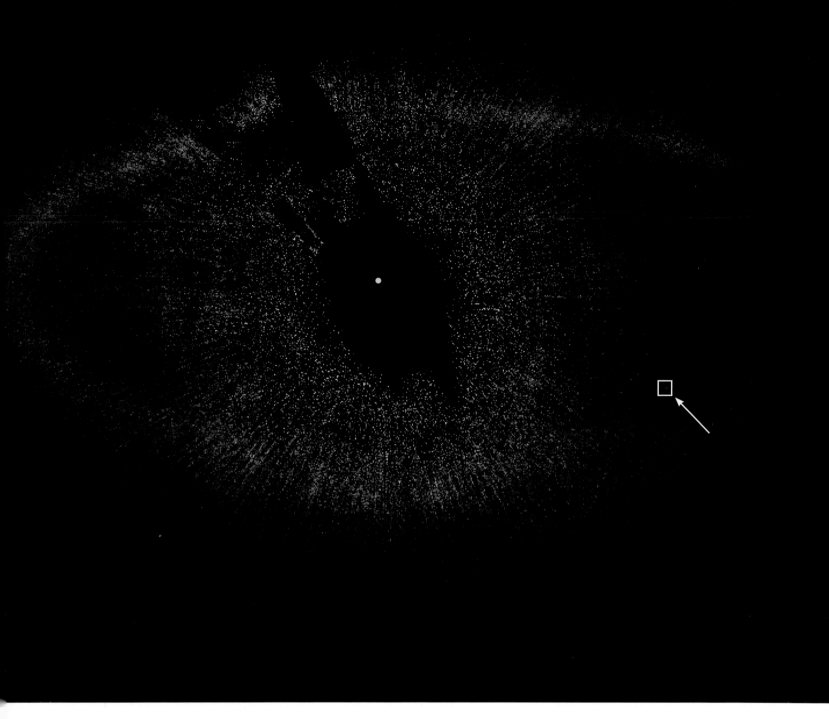

There is a strange population of brown dwarfs unattached to any star. They may be numerous, but their origin is unclear. These lonely objects have also been called 'rogue planets', thrown out of the systems in which they were formed through gravitational interactions, but it seems unlikely the necessary number can be produced in this way.

Thanks to the ever-increasing catalogue of exoplanets, we can be much more confident that Earth-like planets are common in our Galaxy, at least around single stars. In binary systems, it is difficult to see how a small planet could survive for long without being thrown out of the system. Nevertheless there is at least one known exception, in which a large planet has been detected orbiting a Sun-like star in a triple system.

Fascinating as this weird and wonderful collection of planetary systems is, we obviously have a special interest in a particular type of solar system, those that include a small, rocky, and wet planet. Let us focus now on our own newly formed planet, Earth.

▲ First direct sighting of an exoplanet

Taken from the Advanced Camera for Surveys aboard the Hubble Space Telescope, this is the first ever image of a planet from another solar system. The blue dot in the centre represents Fomalhaut. Images taken in 2004 and 2006 showed that Fomalhaut b (a tiny dot in the area marked by the white square and indicated by the arrow), is moving in orbit around its parent star.

9.2 BILLION YEARS A.B. (AFTER THE BANG) TO THE PRESENT DAY (13.7 BILLION YEARS A.B.)

CHAPTER 5 **The Emergence of Life**

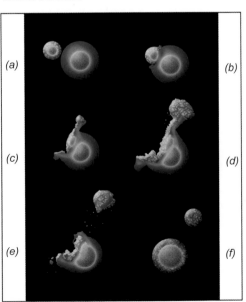

At around 4.6 billion years ago, the Earth had finally formed but it was completely molten, and before the surface could cool a dramatic event took place that resulted in the formation of the Moon. The favoured theory at the present time is that this was due to a giant impact, when the Earth collided with a body, perhaps as large as Mars. The two merged, and debris was scattered to form the Moon. The fact that the Moon is less dense than the Earth shows that the actual cores of the two bodies were not involved in the formation of the Moon but merged to form the Earth's present core.

The rôle of the Moon

Our Moon appears to be special, and has an essential rôle in the evolution of life on our planet. The Moon stabilises the tilt of the Earth's axis, which is currently 23 degrees and does not vary by more than a degree. If the Moon had not been present, this tilt would have varied markedly, and the climatic situation would be very different. Compare this with Mars, which has no comparable satellite; its two attendants, Phobos and Deimos, are so small that their influence is negligible. There is thus no stabilising force, and the tilt of the Martian axis varies from 11 to 35 degrees over a cycle of roughly 100,000 years. The evolution of life depends upon the long-term stability of climate. If Earth's axis of rotation varied wildly over a short period, this stability would be lacking, and life as we know it would not have developed. It seems we owe our grateful thanks to the Moon for making our existence possible.

The most obvious effect of the Moon on the Earth is in the tides that it raises. The friction this causes slows the Earth's rotation, and this process is continuing today. An equally important effect is to increase the separation between the Earth and its satellite; the distance between them is increasing at a rate of 4 centimetres ($1^1/_2$ inches) per year.

As one might expect, the Earth has a similar effect on the Moon, and the Earth's mass is 80 times that of the Moon, so its influence has been even more extensive. Long ago the Moon's rotation was tidally slowed until it became 'captured', or synchronous, meaning that its period of spin became exactly equal to its orbital period. The result is that the Moon always presents the same face to the Earth. It is important to remember that although the Moon keeps the same face turned toward us, it does not keep the same face turned toward the Sun, and the idea that there is a 'Dark Side' to the Moon is completely wrong. Day and night conditions on the Moon are the same in both hemispheres, apart from the fact that the Earth will never be seen from the averted hemisphere.

The speed of the Moon's rotation quickly became a constant, but the speed with which it travels along its elliptical path round the Earth never has. Following the usual traffic laws of the Solar System, the Moon moves fastest when near 'perigee', its closest point to the Earth, and slower elsewhere. Therefore the position in orbit and the amount of rotation become out of step. The result is that, seen from the Earth, the Moon seems to rock to and fro. Sometimes we see a little around the mean Western 'limb', and sometimes a little around the Eastern edge. Altogether, due to this and other smaller 'librations', as these wobbles are called, we can examine from the Earth a total of 59 per cent of the Moon's surface, although never more than 50 per cent at a time. It is only 41 per cent of the surface that we cannot see.

Our planet – cradle for life

At first, the Earth was molten, and far too hot for life to appear. It cooled down gradually

over roughly 500 million years, and formed a solid crust. The original atmosphere was made up largely of hydrogen, but this could not last. The energetic atoms soon leaked away into space, as the Earth's gravitational pull was, and is, too weak to retain them. There may even have been a period when the Earth had no atmosphere at all, but this changed. Volcanic activity would have been much more common – and much more violent – and eruptions from inside the globe soon sent out sufficient quantities of gases to produce a new atmosphere. Of course, this atmosphere was very different from that of today, notably in its lack of oxygen. However, as the atmosphere cooled water began to condense and there followed the period we could call the Great Rains, which lasted for long enough to fill the lower lying regions with water and produce the first oceans.

There was also a period of bombardment from material left over, so to speak, when the planets were formed. This is very evident when we look at the lunar surface, where the craters were produced by bombardment during this period. Of course, the Earth was equally bombarded but erosion has removed most of its scars. It is worth noting that, had it not been for the continuing tectonic activity, the crashing together of plates and the squeezing up of mountains, water would have covered the whole of a smooth globe to this day. The tectonic forces are driven by heat from the decay of uranium and other unstable heavy elements deep within the Earth. These materials must, as we have seen, have come from previous cataclysmic star deaths. So many distant events have been responsible for making it possible for the stage to be set for the emergence of life.

Life emerged much earlier than has often been believed. The first organism able to reproduce itself probably appeared about 4.3 billion years ago. The earliest evidence of life, attributed to the first organisms which were clearly primitive, is a marked rise in levels of oxygen in the atmosphere. The fact that the presence of significant quantities of oxygen is an unmistakeable sign of life gives hope to those scientists developing missions to search for Earth-sized planets around other stars; interstellar travel may be some way off, but we may be able to see the signature of life from a great distance. The oldest evidence of life yet discovered dates from 3.8 billion years ago and is found in ancient rocks from the island of Akilia in Western Greenland.

The exact process by which life arose is still unclear; contrary to popular myth, no one has yet come close to repeating this feat in the laboratory. The theory (unproven) goes that chemical reactions were driven by energy from sources such as lightning strikes and short-wave radiation from the Sun. As time went by, more and more complex molecules were produced, until eventually a molecule appeared which could replicate itself. The ability to replicate, or reproduce, is fundamental to anything that we think of as life. The replication was not perfect; each generation brought with it a chance of random variations – errors in the copying process. Some of these random mutations, as they are called, were

Panspermia

The late Fred Hoyle, and his colleague Chandra Wickramasinghe, building on conjecture by the Swedish scientist Arrenhuis, maintained that comets could dump viruses into the upper atmosphere, causing widespread epidemics. (Viruses are strands of DNA or RNA that use the apparatus of living cells to replicate. Some biologists dispute that they are actually alive in the classic sense.) Again there has been little support for this idea, and the idea has never been taken seriously by medical experts.

Probably the most unusual serious theory in connection with life from space was proposed by no less an eminence than Francis Crick, co-discoverer of the double-helix structure of DNA. With the

▲ **Volcano**

Mount Erebus, the most active volcano in Antarctica, as photographed by Patrick Moore. Earth's early oxygen-free atmosphere is believed to have formed as a result of a period of prolific volcanic activity.

more successful, surviving longer or reproducing more easily than others, and so were more likely to form the next generation. This competition between slightly different forms is at the heart of what has become known as evolution. The long stately process which must have led from these simple replicators, no more than complex molecules, to the vast variety of life that we see around us today, had begun.

The earliest known fossils are of bacteria. These organisms probably lived in the hot oceans found on Earth at this time. We can be fairly confident about their age, because geological methods can tell us the age of the rocks in which the remains of these primitive organisms are found. In rocks of this time we also find what are called stromatolites,

chemist Leslie Orgel, he put forward the theory of 'directed panspermia', according to which life was deliberately sent to Earth by beings from an advanced technological civilization from far across the galaxy. It was pointed out that the chances of micro-organisms being passively transported from world to world across interstellar space were slight, but things would be different if preparations were made. Different types of micro-organism could be carried in a spaceship and deposited here to flourish and develop. When the theory appeared, it is perhaps fair to say that scientists in general were stunned rather than enthusiastic, but ideas of this kind are extremely difficult to disprove.

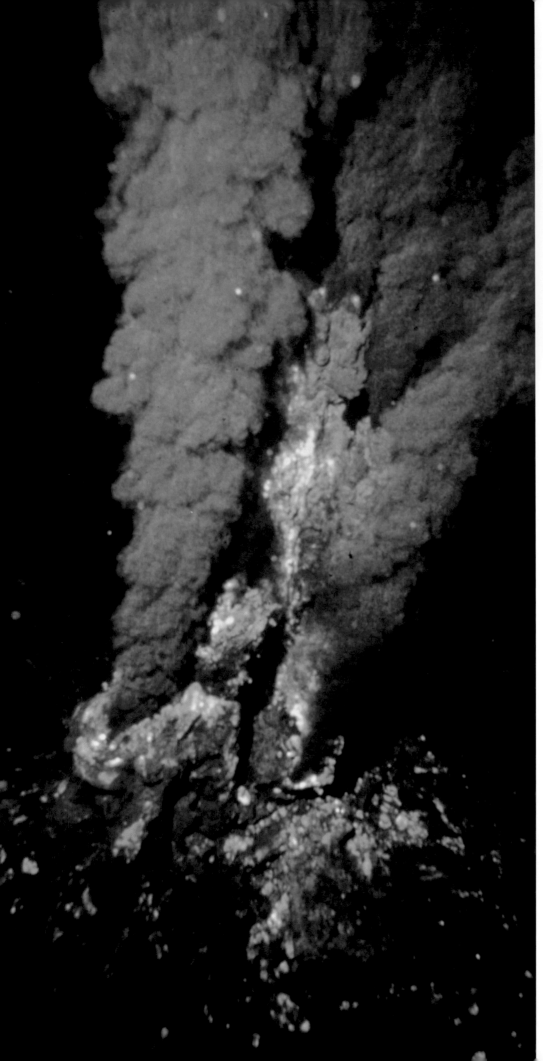

◀ Black smoker

More properly known as hydrothermal vents, black smokers are sites over a mile below the surface of the ocean where superheated gases are escaping from fissures in the Earth's crust. They are found in mid-ocean ridges, and surprisingly a huge variety of organisms surround them, thriving despite high temperatures, a complete absence of sunlight, and high acidity. These creatures form a food chain completely independent of energy from the Sun, but that is instead based on chemicals emerging from the vent itself.

▲ Colony of tubeworms

Relatives of these creatures live near black smokers in extremely hostile conditions. They have no mouth or stomach and survive by absorbing chemicals in the water through their skin. The fish is a vent fish.

rock-like structures built by blue-green algae, also known as cyanobacteria; stromatolites too date back to around 3.5 billion years ago, and, remarkably, some types survive today, notably in parts of the Northern Territory of Australia. During the early stages of Earth's history, cyanobacteria played a major role in producing free oxygen, and this began the process of producing a breathable atmosphere.

We have found that life is extremely versatile; some forms are amazingly tolerant, and can thrive in the most unlikely places. For example, one of the first places life may have appeared is around what are termed hydrothermal vents, often known as black smokers. These are fissures in the floors of the oceans that leak out hot, acid-rich water from below; the effluvia are often black – hence the nickname. The temperature of the water emerging from these fissures, at least a mile under the sea's surface, may be as high as 400 °C. Water is able to reach this temperature, higher than its normal boiling point, due to the pressure, which is 25 times that due to the atmosphere at ground level. Remarkably, the fissures are teeming with specialized life forms such as tubeworms, shrimps and even clams, which survive in an environment as acidic as vinegar that

FOSSIL TRILOBITE
Elrathia kingii
Utah, USA
Cambrian Period
535 Million Years Old

▲ **Inspirational history**

It was a book by Patrick Moore, *The Earth*, in the school library, that introduced Brian to the amazing story of the trilobites, and inspired him to a life-long passion for astronomy. There were once 15,000 species of trilobite, and by the time they were wiped out by the Permian extinction 250 million years ago, they had roamed the Earth for 300 million years. By comparison, we humans have so far been around for less than 200,000 years. The trilobites' nearest living relative is the horse-shoe crab. This particular trilobite recently found its way to a nature shop in New York, for sale along with a fine selection of meteorites, dinosaur bones, and other clues to our distant past.

would be instantly fatal to most other forms of sea life, and without receiving any energy from the Sun.

Worldwide, the fossil record allows us to trace the evolution of living creatures. Generally speaking, life evolved rather slowly; for a long time it was confined to the sea, and only during what is termed the Devonian period, around 400 million years ago, did life spread to the land – plants first, then arthropods (such as insects, spiders and crustaceans) and vertebrates. The plant growth on land continued to produce changes to the composition of the atmosphere. Plants survive using photosynthesis, which removes carbon dioxide from the air and uses it to build food in the form of sugar molecules. A waste product of this process is oxygen, which plants release into the air.

Graveyard of the dinosaurs

The greatest catastrophe in life's history occurred at the end of the era known to geologists as the Permian, 250 million years ago. The Permian lasted for around sixty million years, and seems to have been a time of widespread deserts. Most of the world's land masses were joined together in a vast continent, which has been called Pangaea. It seems that this Permian extinction, often known more poetically as the 'Great Dying', was the greatest in history and wiped out most of life on Earth. This, of course, can be established by the fossil record, but there is no crater left to guide us as to the cause of the disaster. Instead, we must depend on certain carbon molecules known as fullerenes. These molecules form a cage-like structure, most often in the shape of a ball, and inside this cage single unreactive atoms are trapped at the time of formation. The helium and argon found in fullerenes at the end of the Permian seem to have come from space, produced in the atmosphere of a star that exploded as a supernova before the Sun was formed. These chemicals may be the remains of a meteorite that carried material left over from the beginning of the Solar System. It is suggested that, as a consequence of the impact, there may also have been huge amounts of volcanic activity, covering the entire land surface in lava to a depth of three metres (nine feet). It is therefore far from surprising that 90 per cent of all marine species and 70 per cent of land vertebrates failed to survive.

Reptiles began to appear throughout the Permian, and we come to the age of the dinosaurs, some of which were huge and ferocious hunters while others were small and herbivorous (plant eaters). One small harmless dinosaur, no larger than a canary, has been nicknamed the Tweetieosaurus.

Dinosaurs ruled the world for almost 200 million years (in comparison, human beings have been on Earth for less than 200,000 years), but then, at the end of the geological Cretaceous period, 65 million years ago, the great dinosaurs suddenly vanished. Yet the extinction may not have been total; it now seems certain that some of the smaller species survive to the present day in the form of their feathery descendants, the birds. The departure of the dinosaurs may have been a good thing from our point of view, because it meant that mammals could diversify from small shrew-like animals into the wide variety of species we see today. The apes that evolved in the Miocene period (25 to 5 million years ago) are our direct ancestors.

Investigation of the cause of extinctions is a popular pursuit, and opinions vary. So far as the demise of the dinosaurs is concerned, the favoured theory of today is that a large meteorite struck the Earth, throwing up a colossal amount of dust, and causing global devastation – it has even been claimed that the site of the impact has been identified;

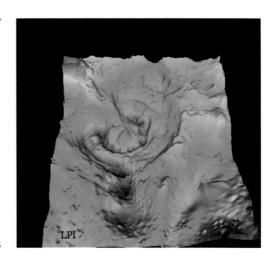

▲ **Chicxulub basin**

The Chicxulub crater in Mexico may be the site of the impact that ended the reign of the dinosaurs.

▲ Exploring the surface of Mars

In this reconstruction, NASA's rover vehicle, Spirit, stands proudly on the surface of Mars, halfway up Husband Hill. This robot geologist, along with its twin, Opportunity, has provided the best evidence yet that there was once liquid water on the planet's surface.

Chicxulub off the coast of Mexico, where we can detect the eroded tracts of a vast crater. The evidence is based mainly on the fact that rocks laid down at this period over a large area contain more than the expected amount of iridium – an element comparatively rare on Earth but characteristic of meteorites. We cannot be certain the impact wiped out the dinosaurs, but the theory is widely supported.

It has been useful to spend some time in discussion of life on Earth, because the question must now be asked whether this sequence of events has been duplicated elsewhere. If, elsewhere, there is an Earth-like planet orbiting a Sun-like star, there have been suggestions that we should expect to find some kind of life, even though we have no idea how life started. But we will never be sure until – if – we detect signals from another

Mars Express

For many years, people have speculated about the possibility of life on Mars, which is the only world in the Solar System that is not too unlike the Earth. It is much smaller, so that it has lost much of the atmosphere it may once have had, and it is colder, because it is around 44 million miles further away from the Sun, but there are no toxic clouds or lethal radiation zones. Results obtained in 2004 by the

Mars Express orbiter, and the two US rovers Spirit and Opportunity, prove that salty seas once covered parts of the surface, and presumably conditions then were suitable for life. The Martian engineers invoked by Percival Lowell a mere hundred years ago to explain the canal-like features he thought he saw on the surface have been banished to the realm of science fiction, although many people,

civilization. The search is on; however, all direct searches for signals from other intelligent beings (SETI – the Search for ExtraTerrestrial Intelligence) have so far drawn a blank.

Is there life on Mars?

What factors do we need to consider when calculating the odds of succeeding in our search? One point must be cleared up at once: we are discussing life as we know it. All life of the kind we can understand is based upon one type of atom, the atom of carbon; only it can link up with enough other atoms to form the complex atom groups or molecules that are needed. It follows that life, wherever it exists – here, on Mars, or on a planet in a distant galaxy – must be carbon-based. Worlds such as the airless Moon must be rejected

including scientists, were sorry to see them leave the stage, but lowly organisms could exist there even now. It is not unreasonable to suppose that if life appears, it will evolve as far as local conditions permit. It may well be that life appeared on Mars and began to evolve but had little opportunity to diversify before conditions deteriorated.

It was recently claimed that evidence of micro-organisms had been found in rocks thought to have been thrown from Mars to Earth by the impact of a huge meteorite on the surface of Mars. However, there is doubt as to the accuracy of the conclusions, and it would take a sample of the Martian surface brought back by a probe, showing clear signs of life, to allow us to conclude that life will appear wherever conditions are favourable.

▼ **Mars Express Spacecraft**

The European Space Agency's Mars Express probe carries a stereo camera that produces three-dimensional images.

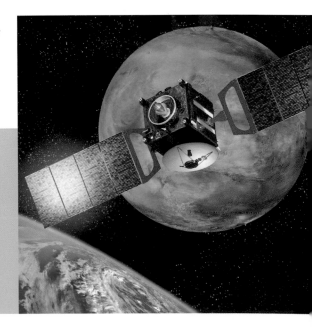

▲ Fact and fiction

Renowned for his scientific writing, Patrick has also written many science-fiction novels.

out of hand. In our Solar System, perhaps only the Earth is suited to complex, intelligent life forms. The counter-argument, of course, is that we may be completely wrong, and that there can be intelligent beings whose bodies are based upon atoms of gold, and who breathe in atmospheres of sulphuric acid. Beings of this kind (BEMs or Bug-Eyed Monsters) are much loved by science-fiction writers, from HG Wells onward, but if they do exist then the whole of our modern science is wrong, and this seems wildly improbable.

We have at least established that many stars do indeed have planetary systems, but for a planet to bear life there are several conditions that must be fulfilled. (Again, let us stress that we are considering only life of the kind we can understand. Once we enter the realm of completely alien life forms, speculation becomes endless and, for now, we intend to restrict the discussion to carbon-based life.) The planet must have an atmosphere that contains sufficient free oxygen; it must have a solid (or possibly liquid) surface; there must be an adequate supply of water; a fairly equable temperature; and a long period over which conditions do not change dramatically. All these conditions are fulfilled by the Earth, but by no other body in the Solar System.

However, there may be other less obvious requirements. For example, a fairly regular alternation of day and night seems advantageous. If one hemisphere of the planet is in permanent darkness, and the other permanently illuminated by the Sun, violent winds would ensue, rain would be impossible and temperatures would hardly be conducive to life; freezing on one side and boiling on the other. Of course there might be a favourable zone on the terminator (the boundary between the dark and light sides).

Let us focus on temperature. Around a star there is a region known as the habitable, or 'Goldilocks' zone or ecosphere, where a planet will be neither too hot nor too cold for life to exist. Neither of the orbits of Venus or Mars lie within the ecosphere; Venus is too close to the Sun, and too hot, and Mars is too far away, and too cold. Only our planet moves comfortably in the middle of the zone; the Earth, like Baby Bear's porridge, is just right. A star less luminous than the Sun will have its ecosphere closer-in; with a more powerful star, the ecosphere will be further out. Many requirements are self evident, and rule out

▶ Christmas on an alien world

One of the playful visions of what life might be like on another planet by Gertrude Moore, Patrick's mother.

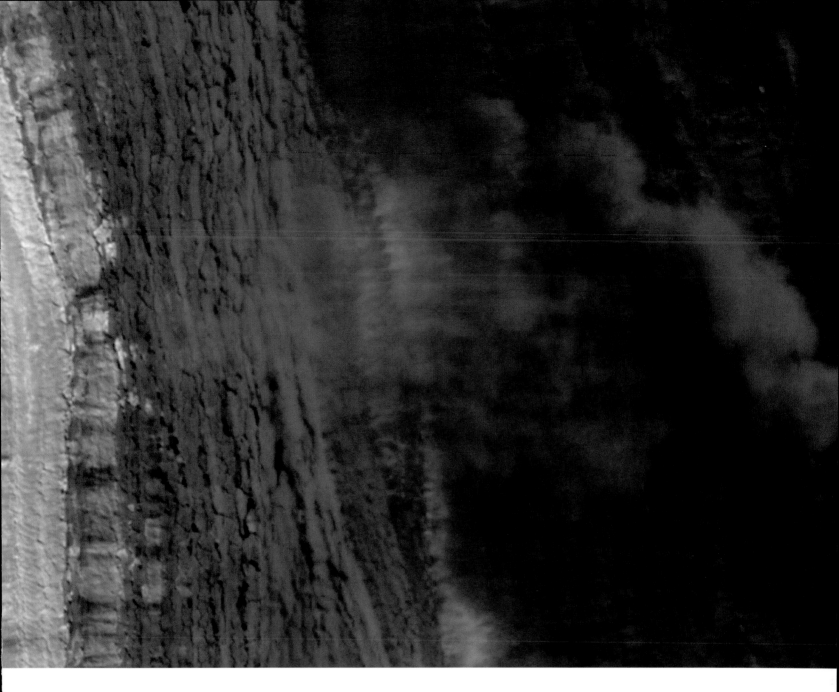

many stars as candidates for planetary systems; a strongly variable star, for example, would give any planet a most changeable climate.

We have already seen that our Galaxy has around a hundred billion stars, an average number for a large galaxy. It seems likely from present day observations that most single stars have planets, leaving us with around forty billion solar systems. How many of these planets lie within their star's ecosphere? On the evidence of our own Solar System, the only one that we have sufficient knowledge of, we might surmise that each solar system possesses one planet within its ecosphere. We should rule out planets around strongly variable stars, however, so we might be left with twenty billion suitably placed planets. How many of these will be rocky? This is a new problem – as we have seen, other solar systems seem to have gas giants in the Goldilocks region. It is difficult to determine the probability of finding a rocky planet in a suitable place, but out of the 120 or so systems known, 30 have no giant planet in the way, so using that fraction as our optimistic estimate we have five billion planets on which we think conditions could allow life to form. On how many of these did it form? To answer this question – perhaps the most difficult of all – we need to know and understand the exact mechanism by which life

▲ **Avalanche on Mars**

This is the incredible view from Mars Reconnaissance Orbiter, directly overhead the sheer cliffs near the North Pole of Mars. The billowing cloud of ice and dust to the right of the image is some 200 metres (650 feet) across. It has plunged from the top of the 700-metre (2,300-foot) cliff. The snow-like material at the top of the cliff is carbon-dioxide frost.

started. It is fair to say that biologists have no experimentally tested detailed theories, and this makes reducing the probability to a single number very difficult. If the probability is even one in a thousand billion, finding just our civililization within our galaxy seems an amazing fluke. If, as some believe, it is closer to one in a hundred, then we will have many millions of promising planets to search.

In our own Solar System the most likely candidate is Mars, which after all is in many ways not too unlike the Earth. It has an atmosphere, admittedly tenuous; its rotation is a mere half-hour or so longer than ours, and the surface temperatures are not intolerable. The main problem is that there is no adequate shield against harmful radiation from space. If there is life, it is most likely to be found beneath the surface, and from this point of view the situation seems rather more promising than it did a few years ago. Moreover, the existence of life in the remote past, even if it has now died out, would be immensely significant.

Mars was certainly once a relatively warm, watery world. The remarkable exploits of the American vehicles Spirit and Opportunity have shown that there used to be extensive seas; Mars was then fully able to support life. Large amounts of water persist in the form of ice in the northern arctic regions, where the Phoenix spacecraft landed in 2008. Phoenix was able to use its robotic arm to sample the ice, and perform chemical analyses. It is possible – and the subject of much debate – that this water might still occasionally gush out, affecting the surface. When astronauts land, it is not impossible that they will discover Martian fossils – the dream of every palaeontologist!

Mars may hold the key to one of the main problems facing us. If we find life any kind, past or present, however lowly, it will show that life will appear wherever conditions are at all suited to it, and will develop as far as its environment allows. We will be entitled then to assume that life is likely to be common throughout the Galaxy. Yet even then our problems will not be over.

Some biologists believe that once life has started, intelligence is inevitable, while others argue equally convincingly that intelligence like ours is a true one-off. How many of these intelligences could we detect? They must have reached or surpassed a level of technology achieved by the human race only in the last hundred years. And then we must wonder how long a civilization capable of communication will last before being destroyed,

▼ Project Ozma

The 84 ft (25.9 m) radio telescope at Green Bank, West Virginia, was used by Frank Drake and his team in 1960 to carry out the first search for extraterrestrial intelligence. It has since been replaced by this 357 ft (110 m) dish.

Tau Ceti and Epsilon Eridani

Two nearby stars seemed particularly good candidates as centres of planetary systems: Tau Ceti and Epsilon Eridani, both of which are easily visible with the naked eye. Both are of solar type, though less luminous than the Sun. Tau Ceti, at a range of 11.9 light-years has 40 per cent of the luminosity of the Sun, while Epsilon Eridani lies 10.7 light-years away and has 30 per cent of solar luminosity. These were the target stars for the first experiment for SETI, carried out by Frank Drake and his team in 1960. The investigators 'listened out' at a wavelength of 21 cm (1420 megahertz), the frequency of emissions sent out from the clouds of cold hydrogen spread throughout the Galaxy, but

the results were negative. Drake formulated the famous 'Drake Equation' to calculate the probability of life existing elsewhere in the Universe. Unfortunately, as Drake himself was the first to point out, the formula contains too many unknowns for firm conclusions to be drawn. The experiment was known officially as Project Ozma, after the fake wizard in Frank Baum's classic children's book, *The Wizard of Oz*.

Tau Ceti has turned out to be something of a disappointment. Like the Sun, it is a yellow dwarf star, its mass is 80 per cent of the Sun's, and its surface temperature is much the same. Its ecosphere should extend between 0.6 and 0.9

◀ Jodrell Bank

The 250 ft (76 m) radio telescope was built at Jodrell Bank by Bernard Lovell in 1957, and has subsequently been involved in SETI projects.

either by natural disaster or by its own folly. In our own case at present, the latter seems more likely. We have reached the point where the uncertainties are biological rather than astronomical, and we await further developments. Now, remember we have so far considered only our own Galaxy – one of billions. The thought that in all of this vastness we could be alone is literally awe-inspiring.

If intelligence exists elsewhere, are there any ways of having a meaningful interaction? We can dismiss modern-type spacecraft without further ado. Even if we could travel at the speed of light, reaching even the nearest star that might be attended by planets would take years, and according to Einstein's theory of relativity, travel at the full speed

astronomical units (1 A.U. is the average distance of the Earth from the Sun), so that if it replaced the Sun in our Solar System, Venus would be comfortably placed. But when Tau Ceti was studied with the James Clerk Maxwell Telescope in Hawaii, the most powerful telescope of its type in the world, things began to look less promising. The star is associated with a disk of debris, and so any planets orbiting Tau Ceti, would be subject to constant bombardment from asteroids of the kind we believe wiped out the dinosaurs. With so many large impacts, it seems unlikely life could have survived.

Epsilon Eridani, on the other hand, is very different in as much as it really is the centre of a

planetary system. In 1998 a dust disk around it was discovered, at about the same distance as that between our Sun and the Kuiper Belt; there are clumps in it, which suggest the presence of a planet. In 2000 a large planet was tracked down by the wobble technique; its mass is rather greater than that of Jupiter, and it has a very eccentric orbit. A third planet has been suspected, further out and much more massive. All these are unsuited to life, but there seems to be no dust close to the star, and it has been suggested that any material there has been swept up by Earth-sized planets. The ecosphere of Eridani lies at a range of just over half the distance between the Earth and the Sun.

▶ **Voyager 2**

Launched in 1977, Voyager 2 flew past Uranus (1986) and Neptune (1989), returning the first ever close-up images of these outer planets. It is the first man-made object to leave the Solar System altogether and head for the stars. It carries gold-plated gramophone discs containing images and sounds from Earth, in case it encounters alien life. The records include a schematic diagram showing Earth's position in the Galaxy, so that any recipients of these interstellar gifts will be able to thank us in person.

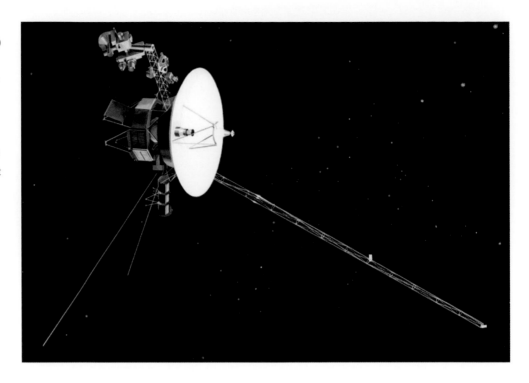

of light would use up an infinite amount of energy – which is another way of saying that it simply cannot be done. Obviously, using our rockets would involve a journey lasting for centuries, and devices such as space-arks, where the original travellers die early in the voyage and only their descendents survive to make 'planetfall', are for now likely to stay as science fiction. Interstellar travel will require a technological breakthrough, which may come tomorrow, within a year, within a century, within a million years – or never. Until it does, from a material point of view we are confined to the Solar System.

So far as communication is concerned, we have as yet tried only one method: radio. Radio waves travel at the same speed as light, and therefore the time of travel between ourselves and the nearest promising stars amounts to only a few years. Moreover, radio communication over a good many light-years would be possible with our present equipment. If there were astronomers living on, say, a planet orbiting the star Tau Ceti, 11 light-years away, they could pick up signals of the strength we are capable of sending. Similarly we could pick up signals from them.

We might expect any artificial transmission to be based upon mathematics; after all we did not invent mathematics – we merely discovered it. Various ingenious systems have been targeted and coded messages have been transmitted, not only to ε (epsilon) Eridani but to many other stars. Contact must be slow; if we send a message to Epsilon Eridani in

The anthropic principle

In modern cosmology there has been an attempt to examine what are known as 'anthropic arguments'. These are based on the so-called anthropic principle, which states that the Universe must be the way it is, because if it were to be any different, we would not be here to observe it! To give a trivial example, if the Universe were the size of an atom, creatures complex enough to be conscious could not exist. Sophisticated versions of this argument have been developed, and used to support the hypothesis that we really are unique. However, it is difficult to see how such theories may be tested.

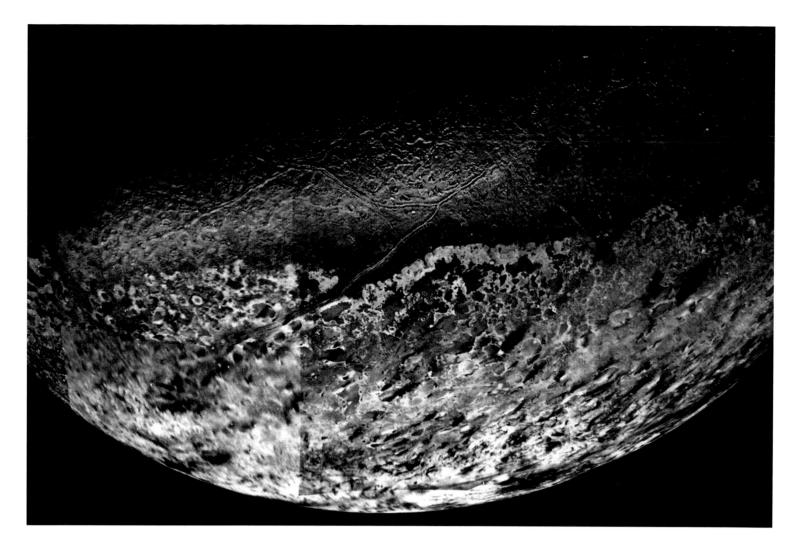

▲ Triton

Triton, Neptune's largest satellite, was the last object to be photographed by Voyager 2 before it left the Solar System. Voyager 2 had travelled 5.6 billion miles in 12 years to reach it.

2009, it will arrive there in 2020, so no reply can be expected before 2031, making quick-fire repartee rather difficult. But it is a measure of our changed attitude that experiments of this kind are considered worth attempting. If no reply is ever received, it may indicate that we are trying the wrong experiment, that there is no technological civilization within reasonable range, or that mankind really is unique.

We cannot send spacecraft to other stars, but a more advanced civilization might well be capable of interstellar travel. We have yet to be convinced by any stories of flying saucers, alien abductions and invaders from Alpha Centauri, but we must remember that ours is a new and no doubt primitive technology. It has been suggested that we should do our best to remain undetected, and even recall the few probes such as Voyager 2 that are now leaving the Solar System permanently, but this would be illogical even if it were practicable (which it is not). Perhaps we should be comforted by the words of Percival Lowell; 'a civilization capable of reaching Earth would have left war far behind and would come in peace'. In any case it is too late for us to 'keep silent'. We have been broadcasting since about 1920, so to anyone within a range of 80 light-years we are 'radio noisy'.

We know that life on Earth has a limited future, and that eventually the increase in the luminosity of the Sun will render our world uninhabitable. We must look ahead, so let us consider the future of the Universe.

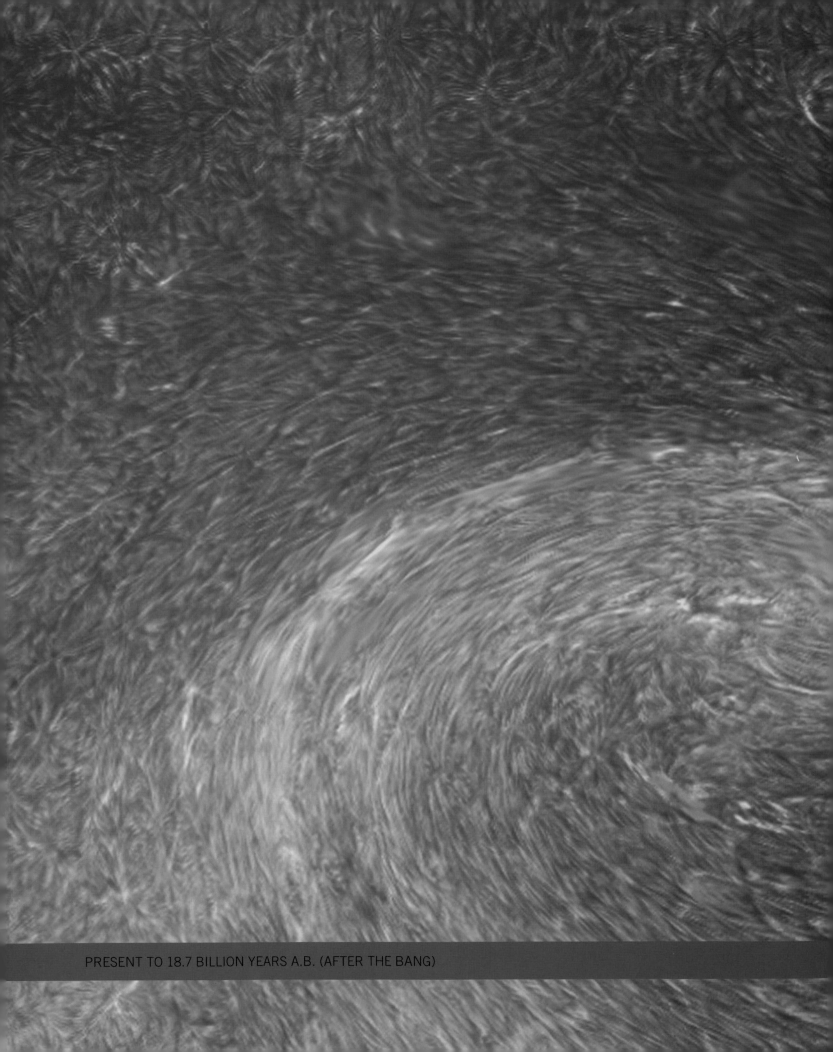
PRESENT TO 18.7 BILLION YEARS A.B. (AFTER THE BANG)

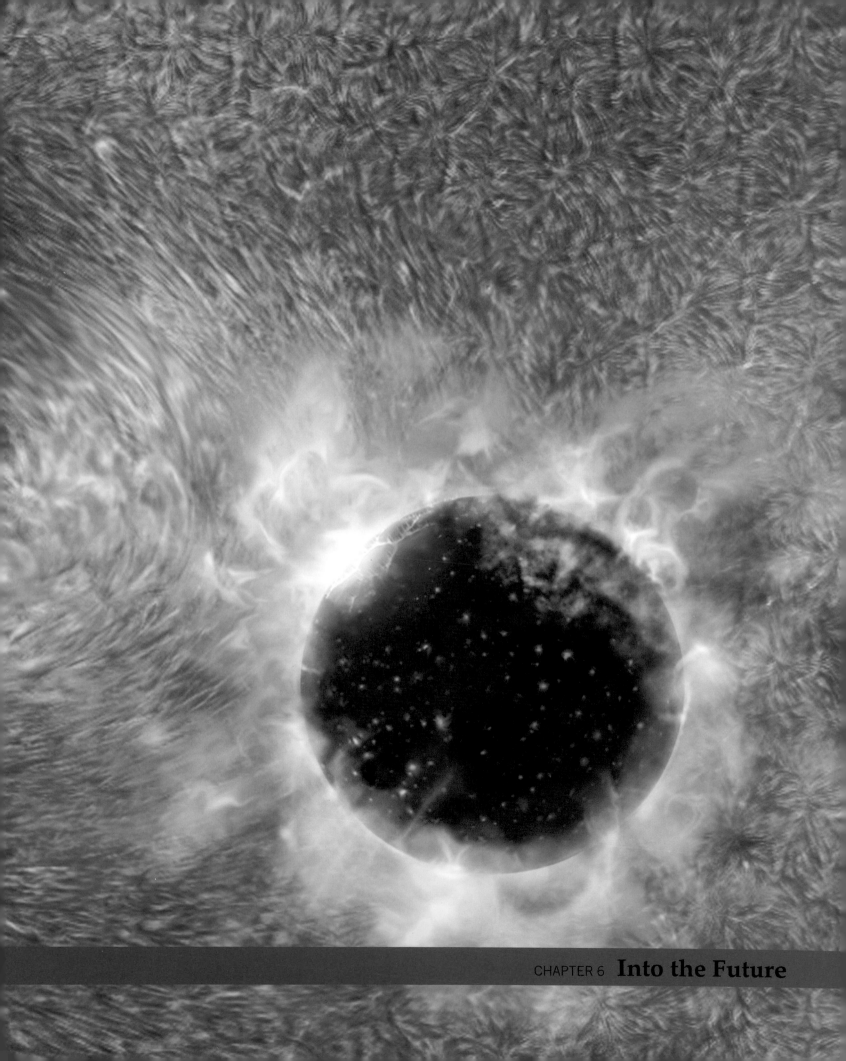

CHAPTER 6 **Into the Future**

When looking into the past we have actual evidence to examine: in the Earth's fossil record we can glimpse the very early stages of our planet's history; in the craters of the Moon we have evidence of ancient cataclysmic meteorite impacts; in the clouds of the Crab Nebula we see the remnant of a violent supernova that occurred almost one thousand years ago. As we gaze at the faint light of the galaxies we are already seeing them as they were millions of years in the past. If we measure the rate at which they are fleeing away from us, we can build up a reliable picture of the state of the Universe as it was billions of years ago; and as we contemplate the cosmic microwave background, we are literally viewing the Universe just 300,000 years after the Big Bang. We can actually see the past.

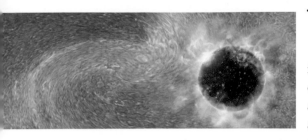

▲ The end is nigh

Five billion years from now, the red-giant Sun will have expanded to such a size that the inner planets Mercury and Venus will be subsumed, and Earth will suffer a fiery death.

▶ One of the wonders of nature

Patrick Moore, standing in Meteor Crater near Flagstaff in Arizona. Caused by the impact of a meteorite around 50,000 years ago, Meteor Crater is about 4000 ft (1200 m) in diameter and 570 ft (170 m) deep. Patrick visited for *The Sky at Night* during the 1970s.

▼ Iron meteorite

Brought from China, the meteorite fell in 1516, during the Ming Dynasty.

The future is more problematic; we cannot see stars and galaxies as they will be in the future, and we have to rely on deduction, mixed with a good deal of scientific speculation. Though many pages of the history of the Universe have yet to be deciphered, we know much more about the Universe of six billion years ago than we do about the Universe of six billion years hence.

The Earth may be insignificant in the Universe, but to us it is obviously of paramount importance, so let us look first at what may lie in store for our own planet. On average, the Earth is hit every few hundred thousand years by an asteroid large enough to cause widespread devastation. Indeed, recently we have tracked several asteroids that have passed alarmingly near the Earth; a few have brushed by at a distance of only a few tens of thousands of miles, well within the orbit of the Moon. They are classed as PHAs or Potentially Hazardous Asteroids, and any one of these is capable of causing another 'Great Dying' if it scored a direct hit. If a PHA were to be seen well before it was due to collide with the Earth, we might be able to do something about it – perhaps by detonating a nuclear device close to it, diverting it from its collision course.

But we have to admit that a collision with a body only a few miles across would be disastrous for humanity, and we might not cope any better than the dinosaurs did. Disturbingly, in spite of efforts to detect just this sort of threat, several of the recent near misses were detected only when they had already passed by the Earth.

There are other wholly possible natural scenarios in which life on Earth could come to a premature end. Geologists have recently become aware of the potential eruption of supervolcanoes, which could result from vast reservoirs of magma under extreme pressure, one of which has been discovered under the Yellowstone National Park in Wyoming. The eruption of any of these could result in a planet-wide cloud of debris in the atmosphere,

▼ **Sumatran supervolcano?**

One of the largest known volcanic eruptions occurred 74,000 years ago when the Toba volcano erupted in Sumatra, leaving the approximately 1000 square mile (3000 square km) Toba caldera behind. This giant depression formed after the collapse of the volcano's cone, seen in this satellite image (immediately below) and at ground level (bottom). The island in the crater lake is a resurgent dome, where magma is active in a chamber below the surface.

▶ Ice on Mars

Three images from the Hubble Space Telescope show how the ice cap fluctuates as the seasons change – here the progression (from left to right) is from Martian autumn to spring and then summer.

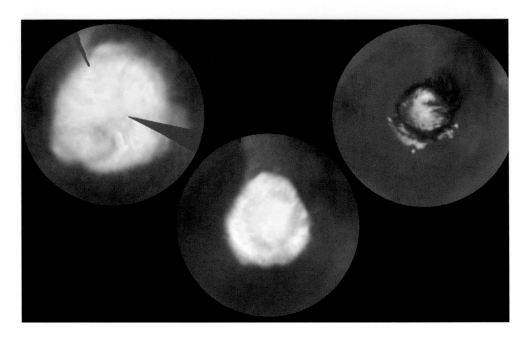

so dense and persistent that most plant and animal life would die from lack of sunlight. It is now thought that some past extinctions may have been due to supervolcanoes.

A man-made disaster is also possible. We now have the capability to destroy ourselves, and we may lack the civilization not to do so. Whatever happens though, the ultimate fate of the Earth lies with the Sun. It is to the Sun that we owe our existence, and it is this same Sun that will destroy our planet.

The end of life on Earth

The Sun is using up its nuclear fuel but, surprisingly, becoming more luminous. This happens very slowly – imperceptibly, as far as we are concerned. As the hydrogen in the centre of the star is used up, so the Sun contracts a little, putting more pressure on the core and raising its temperature. The rate at which the reactions proceed depends strongly on the core temperature, and so fuel is used up faster. A billion years from now, the Sun

▼ Arctic sea ice

The extent of the Earth's north polar sea ice has recently been decreasing at a rate of nine per cent per decade. This picture shows the situation in 2004.

The Sun and the Earth

All through the history of the Earth there have been warm and cold periods. The last Ice Age ended a mere 10,000 years ago; there have been minor variations since then – during the 'Little Ice Age', between around 1645 and 1715, there were times when the River Thames in London regularly froze in winter, and frost fairs were held on it. So what causes these changes in climate?

The Serbian engineer and mathematician Milutin Milanković attributed them to the movements of the Earth itself. The pattern of our orbit round the Sun is not circular; it is elliptical, and the eccentricity of the ellipse varies between certain limits over a

period of just over 400,000 years. At present the Earth's axial tilt (the obliquity) is 23.4° to the orbital plane, which is why we have our seasons, but over a period of around 41,000 years the obliquity ranges between 22.1° and 24.5°; it is decreasing, and will reach its minimum value around the year AD10,000. Precession (the change in direction of the Earth's axis of rotation relative to the stars) also varies, over a period of 26,000 years, and this affects the positions of the celestial poles; when the Pyramids were being built, the north pole star was not Polaris, but Thuban in the constellation of Draco. Taking all these various considerations

◄ Tunguska

In June 1908 something exploded above the Stony Tunguska river in Siberia. Witnesses reported seeing a bright fireball. The explosion was heard 600 miles (960 km) away; 380 square miles (970 square km) of forest were flattened; trees up to 30 miles (48 km) away were felled by the shockwave. It is probable that the explosion was caused by a meteorite 160 feet (49 metres) in diameter entering the atmosphere and vaporizing five miles (8 km) above the ground.

will be powerful enough to give Earth an uncomfortably torrid climate; its inhabitants may well have to abandon the equatorial regions altogether, and huddle near the poles.

But this will only provide a temporary escape. The deserts will expand as the lower latitudes become uninhabitable, and the land available for cultivation of crops will become scarce indeed. The shifting of the continental plates will have long since destroyed the familiar shapes of the continents. Any remaining ice caps will melt, causing an enormous rise in sea level; much of the remaining land will be flooded.

The relentless heat will increase; by three billion years in the future a critical point will have been reached. The Sun will be 40 per cent brighter than it is now, so that all surface water on the Earth will have evaporated; the oceans will be gone, and our world will have become a very hostile place.

If humanity still exists on Earth when the changes in the environment become obvious, how will our remote descendants react? The onset of these changes will be detectable, and

into account, it was suggested that the 'Milankovič cycle' could explain the warm and cold spells.

Other investigators disagreed. After all, we depend entirely upon the radiation we receive from the Sun and although the Sun is a steady, well-behaved star, fortunately for us, it is variable to some extent. The 11-year solar cycle is well known, but there are other factors too. During the 'Little Ice Age' the cycle was apparently suspended, and there were few, if any, sunspots. When the spots returned, there was a period of global warming. Unfortunately our reliable sunspot records only go back a couple of centuries, but the link with

our climate changes cannot be doubted. When the Sun is at its least active, greater numbers of cosmic rays can reach our atmosphere, causing increased cloudiness and a fall in temperature.

Eventually, of course, the Sun will run short of its hydrogen 'fuel' and become a red giant star, with disastrous effects on its planets, but this crisis will not be upon us for so long that it need not concern us. In the foreseeable future we must certainly expect warm and colder periods, but there is every reason to believe that the Earth will remain habitable for at least the next thousand million years.

▶ **Liquid Lakes on Titan**

In the images of Titan's surface from the Cassini probe, some areas reflected very little radar, and the most likely current explanation is that the region is dotted with lakes, composed of liquid methane. This map spans about 150 km. Apart from Mars, Titan is the only other body in the Solar System to possess liquids on the surface.

alarm bells will ring – but it seems unlikely that even a highly advanced civilization could control the Sun. No doubt a committee meeting would be called, but what would be on the agenda? To move the Earth out to a safe distance might be possible, but even this would not provide a permanent solution, as we shall see. It might be possible to remove the Earth from the Solar System altogether and somehow make it self-sufficient, so that it could survive without a Sun. If this proved too difficult the human race might consider mass migration to another world – in another solar system – or else the construction of an enormous, self-supporting space station to accommodate the survivors.

If nothing can be done, as time passes it seems likely that the entire Earth will become a molten, seething mass of magma. There can then be no reprieve; ultimately all life will be wiped out. So much for Earth. In the rest of the Solar System, things may become temporarily more promising for life. Mars will be much warmer than it is now, and its massive ice caps (composed of both carbon dioxide and water) will begin to melt. An atmosphere will develop, and for a short while – a few tens of millions of years or so – Mars will briefly be a hospitable place. However, this situation cannot last for long. Mars is simply too small, and has too weak a gravitational pull to retain its newly found atmosphere for long.

It has been suggested that humanity might find a refuge on Titan, the largest satellite of Saturn, which has a thick, nitrogen-rich atmosphere. Alas, this is not so. Titan has a low escape velocity, and retains its atmosphere only because it is so cold; at a low temperature,

◀ **Betelgeux**

This image of Betelgeux was the first direct picture of the surface of a star other than our Sun. There are some unexpected features, for example the hot spot below the centre.

▼ **Swallowed by the Sun**

The Sun will swell to such a size that Mercury and Venus are consumed. While the current orbit of Earth will be within the red giant, the star's loss of mass will cause Earth to swing outward and so escape. By then, life on Earth will long have ceased to exist. The diagram below shows the size of the swollen red-giant Sun, compared to the size of the inner Solar System today.

gas molecules are sluggish. Raise the temperature by only a few degrees, and the whole of Titan's atmosphere will escape.

During the following half a billion years the Sun will swell to over twice its present size, and although the surface temperature will fall, its luminosity will double. There will also be effects on the Earth's orbit. The Sun's stellar wind will increase in power and our star will begin to lose mass as it evolves into a red giant. This loss of mass means that the Sun's gravitational pull will be weakened and, in response, planets will start to move outward; the Earth will swing out to a distance of around 120 million miles (200 million km) – not nearly far enough for it to escape from the intense heat of the now massively-swollen Sun.

The red-giant Sun

Moving further into the future, at about five billion years from the present day, hydrogen 'burning' in the Sun's core will cease; there will be no hydrogen left – it will all have been converted to helium in the process of nuclear fusion. The core suddenly will no longer be supported by the pressure of radiation emitted from nuclear reactions. Gravitational collapse cannot be prevented; the outer material will rush in, compressing the core and heating the material. Until this point, helium nuclei had been unable to participate in nuclear reactions. In a matter of seconds, however, the temperature will become high enough for another level of fusion to occur. The helium nuclei combine to form beryllium and lithium. This is a much

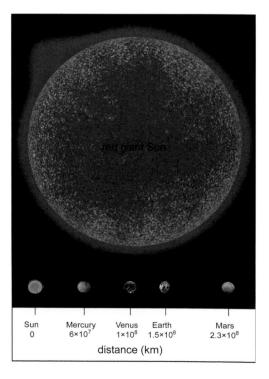

Sun	Mercury	Venus	Earth	Mars
0	6×10^7	1×10^8	1.5×10^8	2.3×10^8

distance (km)

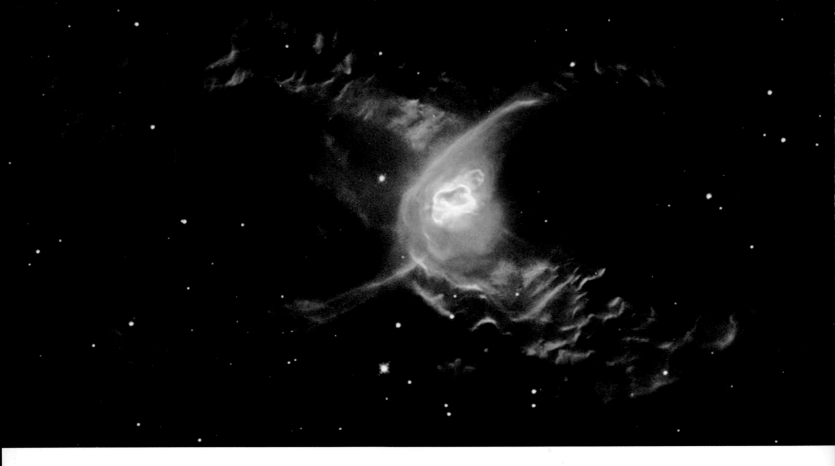

▲ Red Spider Nebula (NGC 6532)

The tangled web of this planetary nebula is the result of a Sun-like star ejecting its gases and becoming a white dwarf. In this case the white dwarf is one of the hottest ever observed.

▶▶ Ring Nebula (M57)

This celebrated planetary nebula has the appearance of a smoke-ring centred around the remains of the ageing star explosion that produced it. If we could see it in three dimensions, our view would be straight down the axis of a tube.

more efficient reaction; the Sun will radiate over 2000 times as fiercely as it does now and it will balloon out to such an extent that Mercury and Venus will be swallowed up. The Sun will have, eventually, become a red giant.

At some stage in its evolution, the ageing red-giant Sun will become increasingly unstable. Its outer envelope will be blown to a distance from the main star by a series of violent pulsations, forming what is known as a planetary nebula.

It is worth noting that a planetary nebula has nothing to do with planets, but is simply the discarded outer envelope of a highly-evolved star. These are the butterflies of the Universe, with many beautiful and varied forms but with lives of only a few tens of thousands of years. The most famous of these objects, the Ring Nebula in Lyra, is easy to locate even in a small telescope, because it is midway between two naked-eye stars, β (beta) and γ (gamma) Lyrae, close to the brilliant Vega; even moderately powerful

The future of the Moon

The Moon will remain linked with the Earth – there is no reason to suppose otherwise – but its orbit will change. At the moment it is moving away from the Earth at the rate of about 4 centimetres ($1^1/_2$ inches) per year, because of tidal effects. The crux of the matter is what is termed angular momentum. The angular momentum of a moving body is obtained by multiplying together its mass, the square of its distance from the centre of motion, and the speed around its orbit – that is to say, the rate of axial rotation. As we have seen, the Moon's axial rotation is the same as its orbital period (27.3 days), which is why it always keeps the same face turned toward us (all the large satellites in the Solar System behave in the same way with respect to their primary planets). Angular momentum can never be destroyed; it can only be transferred. If the rate of axial rotation is slowed down, as happened early in the Earth-Moon system, something else has to increase, and this 'something' is the distance

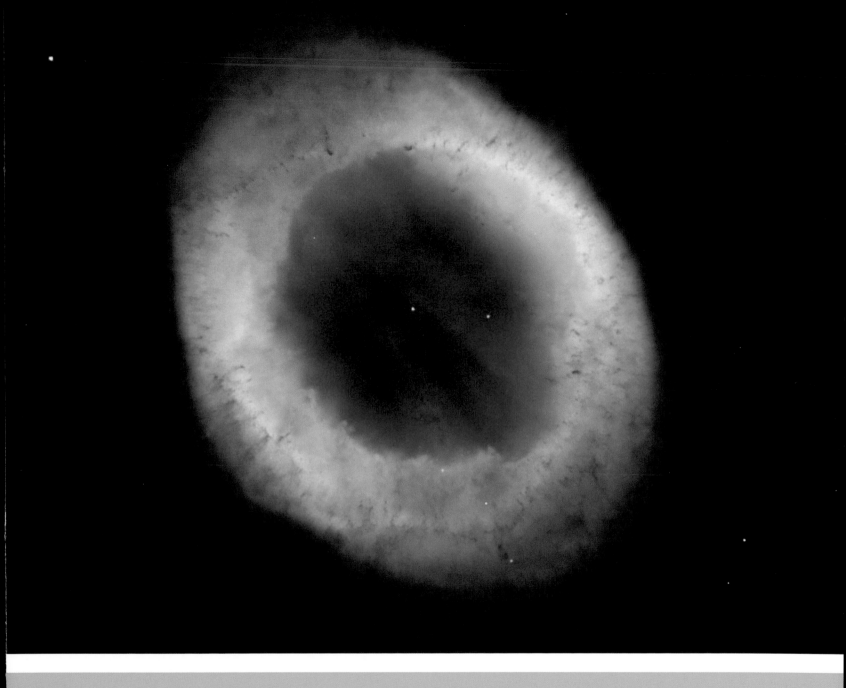

between the two bodies. The situation is rather like that of an ice skater, mid-spin. When she brings her arms in to her sides, angular momentum must be conserved and so she speeds up.

The process is not complete even now, because the Earth's rotation is still being braked by the pull of the Moon, and each day is 0.00000002 seconds longer than its predecessor, though there are also irregular fluctuations not connected with the Moon. It is these that are responsible for the occasional leap seconds that are added and subtracted from the official time. However, the Moon could not go on receding indefinitely. If it moved out to 350,000 miles it would start to draw inward again, because of tidal effects due to the Sun: its orbital period and the Earth's axial rotation period would then be equal, 47 times as long as our present day. If our world survives the Sun's red giant stage this may actually happen, but of course not until long after all life has vanished from the Earth.

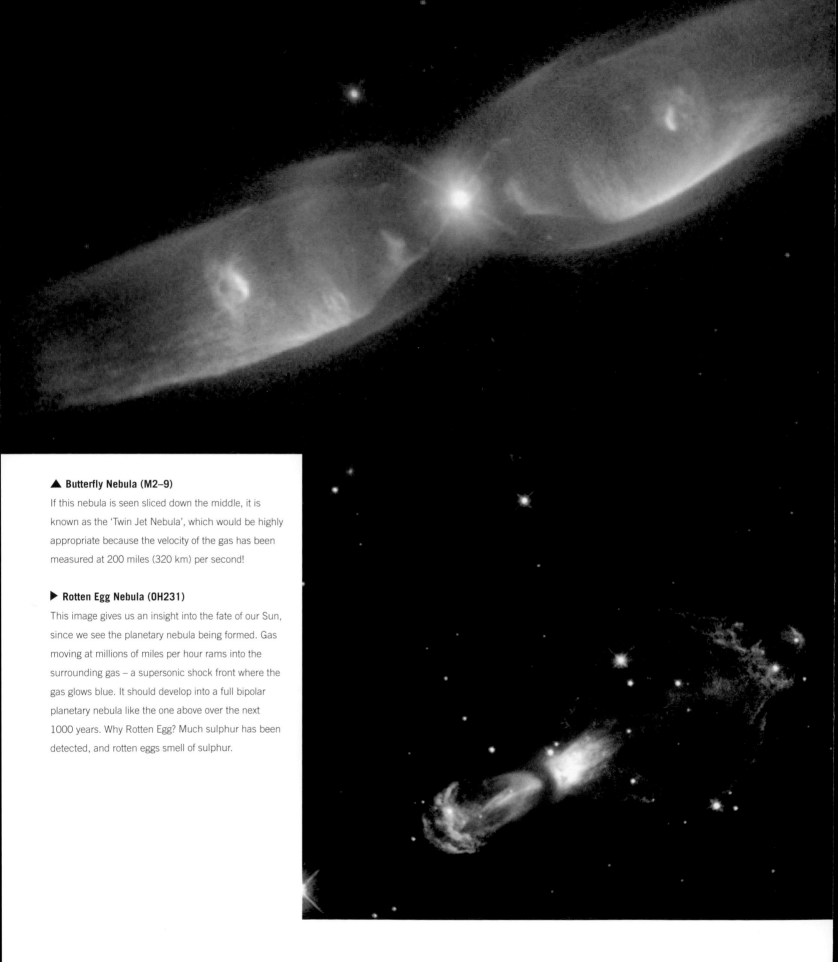

▲ Butterfly Nebula (M2–9)

If this nebula is seen sliced down the middle, it is known as the 'Twin Jet Nebula', which would be highly appropriate because the velocity of the gas has been measured at 200 miles (320 km) per second!

▶ Rotten Egg Nebula (OH231)

This image gives us an insight into the fate of our Sun, since we see the planetary nebula being formed. Gas moving at millions of miles per hour rams into the surrounding gas – a supersonic shock front where the gas glows blue. It should develop into a full bipolar planetary nebula like the one above over the next 1000 years. Why Rotten Egg? Much sulphur has been detected, and rotten eggs smell of sulphur.

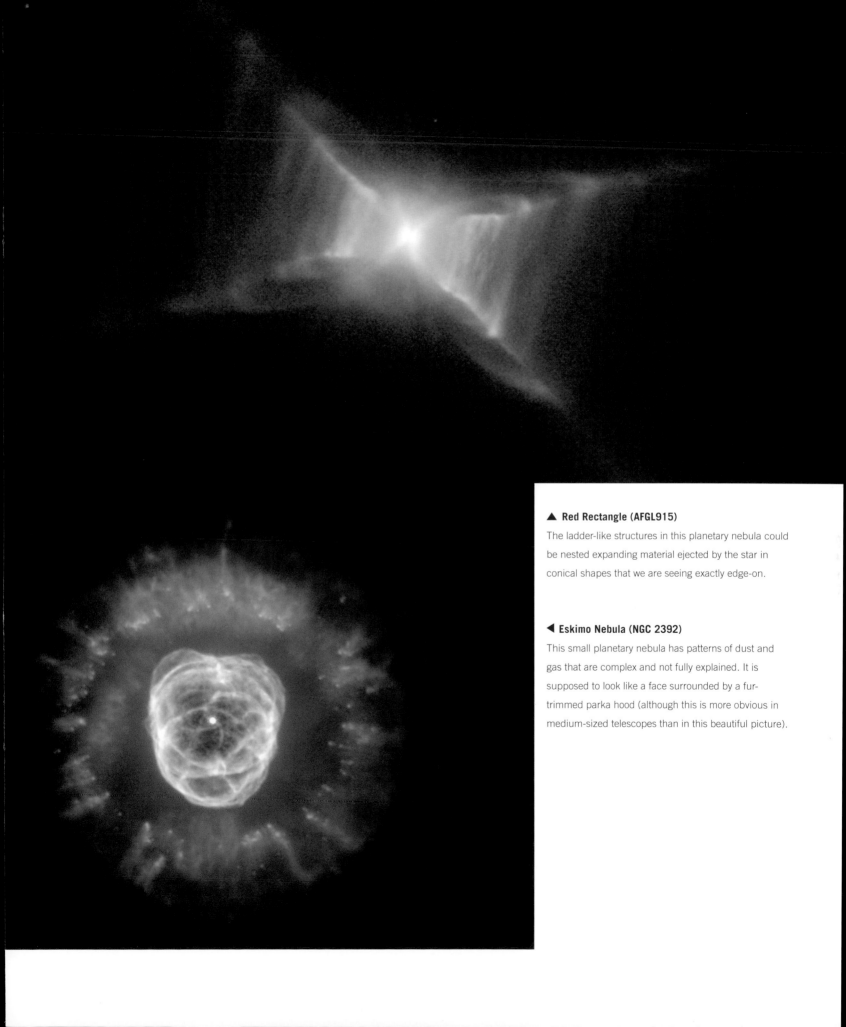

▲ Red Rectangle (AFGL915)

The ladder-like structures in this planetary nebula could be nested expanding material ejected by the star in conical shapes that we are seeing exactly edge-on.

◀ Eskimo Nebula (NGC 2392)

This small planetary nebula has patterns of dust and gas that are complex and not fully explained. It is supposed to look like a face surrounded by a fur-trimmed parka hood (although this is more obvious in medium-sized telescopes than in this beautiful picture).

▶ **An anomalous supernova?**

When supernova 1987a blew up, according to the theory explained here, a neutron star or black hole ought to have been left at the centre of the expanding ring. As yet, no evidence has been found of either.

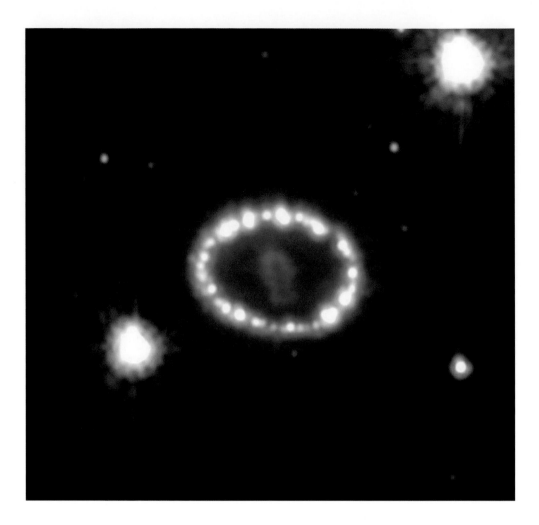

▼ **Bright white dwarf**

Also known as the Dog Star, Sirius is the brightest star in northern skies. Telescopes reveal that it is in fact two stars. In optical telescopes the brightest star is Sirius A, while Sirius B, a white dwarf is 10,000 times dimmer. However, when seen in X-rays, as below, the situation is reversed, and the white dwarf is the strong emitter.

binoculars will show it. In a telescope it looks rather like a dimly luminous cycle tyre. M57 looks symmetrical, but other planetaries show an amazing range of shapes, which must depend on the exact processes by which material is ejected from the central star; it seems that the most common is an hourglass shape, with most material being directed along the axes of the star's magnetic field. According to this model, the planetary nebula appears either as an hourglass or as a ring, depending on whether we are looking at it edge- or face-on. This picture seems to be accurate on the broadest scales, but much of the detail is more difficult to account for. Planetary nebulae are among the most chemically interesting regions of our Universe, with many molecules being produced by reactions driven by the light of the central star in the early stages of the nebula's formation.

White dwarf – the bankrupt Sun

At the same time, back at the central star, now that the fuel available is exhausted, there will no longer be anything to prevent our star collapsing under its own gravity, and this collapse will proceed rapidly. Eventually, the density will become so great that a new resisting force, degeneracy pressure, will begin to work against gravity. Degeneracy pressure is a consequence of the exclusion principle, a fundamental axiom of the theory of quantum mechanics, which holds that no two particles can ever be in the same state – that is to say, if two particles with identical charge, mass and energy come too close together then they will start to repel each other. The star will shrink until degeneracy pressure exactly balances the crushing force of gravity, and at this point the collapse ceases. The new stable state is an incredibly dense core no larger than the Earth, known as a white dwarf. A single teaspoonful of white dwarf material would weigh several tonnes. By now the Earth will have withdrawn to a distance of 170 million miles (270 million km) from the exhausted feeble remnant of the Sun.

What lies ahead? The answer must be 'very little'. The white dwarf is bankrupt; it has no energy reserve, and all it can do is to shine very dimly as it cools, eventually reaching the ambient temperature. The transition to a cold, inert, dead black dwarf takes an unimaginably long time – in fact, it may be that the Universe is not yet old enough for any black dwarfs to have been formed – it seems likely that our Sun will end its life as a tiny, dead star still orbited by the ghosts of its remaining planets.

Neutron stars and black holes

Larger stars meet a different fate. In particular, when the star is so large that the core forming a white dwarf has a mass greater than the so-called Chandrasekhar mass, 1.4 times the mass of the Sun, even the quantum effect of degeneracy pressure is not sufficient to halt the collapse. Instead, the pressure is so great that individual protons and electrons do not survive. Forced to combine, they form neutrons, and we are left with what is known as a neutron star, the density of which far exceeds even that of a white dwarf – a single sugar cube of neutron star material would weigh the same as all of humanity! Neutron stars are extremely small, no more than 15 miles across, but on average are one and a half times as massive as the Sun. If you could stand on the surface of a neutron star your weight would be of the order of 10 billion tonnes. The neutron star is actually the most common form of supernova remnant. We observe them in the guise of mysterious objects called pulsars.

In supernova events in very large stars, even a neutron star is not the end of the line in the rapid shrinking of the core. Once all its nuclear reserves have been used up, the collapse starts, but this time it is so catastrophic that nothing can stop it. The monster star goes on shrinking and shrinking and becoming denser and denser, passing through the neutron-star stage. As this happens, the escape velocity goes up. Any star with less than about eight times the mass of the Sun will end its life as either a white dwarf or a neutron star. If the star is more massive than this, the collapse is literally unstoppable and, as we have already seen, a black hole will form.

Pulsars

Pulsars are rapidly spinning neutron stars, which we see as pulsating sources of radio waves, with several pulses arriving each second. We have already discussed the role of angular momentum in planet formation, and it is important here, too. As the material of the star collapses to form the neutron star, it carries its angular momentum with it, and just as the ice skater bringing his arms into his side speeds up, so the forming neutron star spins faster and faster. Once the collapse is complete, the pulsar will spin at a roughly constant rate. Many pulsars that spin thousands of times a second are now known. Most of these are young; the neutron stars will gradually slow down over time.

What causes the pulses? The emission from material around the neutron star is channelled into narrow beams near the poles of the object. As the star rotates, so these beams flash across the Earth like a lighthouse beam crosses momentarily over a ship far

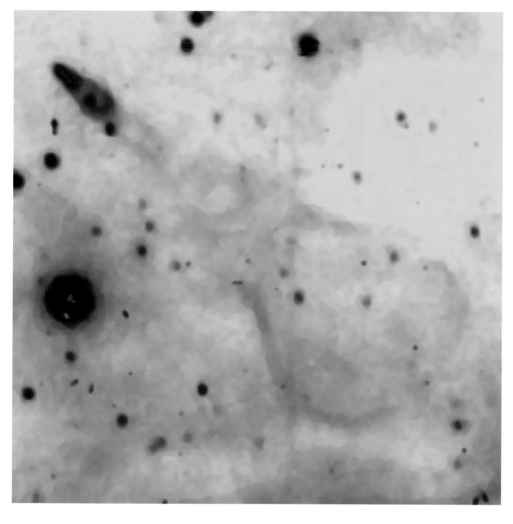

◀ **Guitar Nebula (WNJ2225)**

The wake left by a neutron star travelling through space at about 1000 miles (1600 km) per second created this extraordinary cosmic guitar in the interstellar medium.

out at sea or a watcher on the shore. When the beam is pointing toward us, our telescopes detect a pulse.

Pulsars are the most accurate clocks in the Universe; there are occasionally glitches due to some poorly-understood processes deep in the star, but apart from these rare events and the slowing-down over long timescales, they keep perfect time. They thus provide unique laboratories for astronomers. In particular there is a rare system known as the double pulsar, about which we will have more to say later. There have been reports of the presence of planets orbiting pulsars, the suggestion being that these planets are responsible for tiny changes in the timing of the pulses. However, it is difficult to see how planets could have survived the explosion that accompanied the pulsar's birth.

Remember, we have been discussing the evolution of the core of the star, but something more dramatic is happening outside. As the collapse is suddenly halted, the outer envelope rebounds in a stupendous release of energy. It has become a supernova.

Star worlds in collision

As our Sun grows old, so too throughout the Universe old stars will die and new stars will be formed. Galaxies also are evolving and moving. Our Local Group of galaxies contains only three really major star systems; the Andromeda Spiral, the Triangulum Spiral and

◀◀ **Barrel-shaped nebula**

This beautiful supernova remnant in the Milky Way has a secret that is revealed when viewed in the X-ray part of the spectrum, as here, in false colour. The brilliant blue band may be the remains of a gamma-ray burst, one of nature's most powerful explosions of energy.

▶▶ The Antennae (NGC 4038 & 4039)

The cores of these two colliding galaxies are the orange blobs, and a wide band of chaotic dust stretches between them. Named for their resemblance to the antennae of an insect, the similarity is far less marked in this wonderful image from the Hubble Space Telescope than it is with a less sophisticated ground-based instrument. Eventually the two galaxies will end their cosmic dance and merge, but for now they both shine brightly. This is mostly due to a burst of star formation triggered by the collision.

▼ Arp 299

This pair of colliding galaxies may be the best place to look for a new supernova explosion. A super star cluster in Arp 299 saw its peak of star formation 6 to 8 million years ago, and many stars are now ending their lives in supernova explosions. Four supernovae have been seen since 1990!

our own Milky Way Galaxy. Of these, Andromeda is the largest and Triangulum the smallest. Andromeda, at a distance of between two and three million light-years, is also the nearest and, caught by the mutual gravitational attraction between it and our Galaxy, is approaching us at a rate of 190 miles (300 km) per second. In three billion years time, therefore, in our part of the Universe something really dramatic will occur: a collision between two large galaxies.

If a small galaxy collides with a much larger one, it is simply absorbed and will usually lose its separate identity completely; in any case, it is bound to be severely disrupted by tidal forces; its stars will be literally stripped from it every time it goes near the larger galaxy. Things are very different when two major galaxies collide.

Perhaps it is best to say at this point that although we talk about collisions between galaxies, we do not mean to imply that individual stars might collide. The space between them – remember the Sun is more than four light-years away from its nearest neighbour, Proxima Centuri – is simply too vast and stellar collisions will remain extremely rare, even in the chaotic environment of a merger between two galaxies.

The collision will take several billions of years. If computer simulations are to be trusted, Andromeda will first swing past our Galaxy, and to any watchers present the tiny patch of light would become larger and larger until it came to dominate the night sky as the main interactions begin. As the reservoirs of gas in each galaxy collide, the resulting shock waves trigger the formation of many thousands of new stars, and many of these will be in brilliant clusters dominated by the hot, blue stars. The creation of many massive and therefore short-lived stars means that supernovae will be common, and the shock waves from their explosions will trigger further massive bouts of star formation. The sky will be littered with clouds of glowing gas and dust. After swinging by, what remains of Andromeda will take perhaps 100 million years to describe a stately U-turn, before plunging headlong back into the heart of what was once the Milky Way. Much of the material will be left behind in long streamers, but over time these too will fall into the centre and a large elliptical galaxy seems the likely result. Eventually the black hole at the centre of our Galaxy may well merge with the black hole which almost certainly lies at the

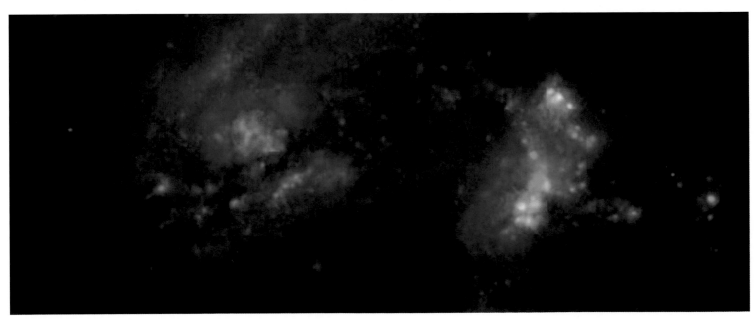

▲ The Mice (NGC 4676)

Three hundred million light-years away, in the
constellation Coma Berenices, the cosmic capers of
the pair of colliding galaxies known as the Mice will
ultimately end with the pair merging into a single giant
galaxy. They are nearer to completing this merger than
are the Antennae, and are classified as a single system
in the *New General Catalogue* (NGC).

heart of Andromeda. It is generally believed that two black holes colliding will combine to
form a single, more massive black hole. Intense radiation is bound to be released, along
with what are called gravitational waves.

Gravitational waves

Gravitational waves are predicted by Einstein's theory of relativity, and can be thought
of as ripples in space itself. They are produced in significant quantities only by the most
energetic of events. However, even then the effect must be small, and gravitational waves
have not yet been detected. Many attempts have been made, but to detect the effect as
space ripples around us, formidable precision is required – equivalent to measuring the
length of a mile-long rod to an accuracy of less than the size of a single atomic nucleus.
Perhaps the best hope lies in using satellites, and various projects are being planned.
Detecting gravitational waves would enable us to probe a whole new set of situations
and objects, including some of the rarest phenomena in the Universe.

Although we have not yet detected gravitational waves, there is compelling evidence for
their existence in the form of a system, unique in our experience and known as the double
pulsar, in which two compact neutron stars orbit each other. Since these remarkable
objects emit extremely regular pulses of energy that can be seen across huge distances,
we are able to time their orbits with great accuracy. Astronomers have discovered that
these two pulsars are spiralling in towards each other, which means that energy must
be being lost from the system; the amount being lost corresponds quite well to the

predicted energy that would go into gravitational waves – but until we detect the waves themselves, we cannot be sure that we have the answer.

The end?

Whatever happens to the central black holes, by this time the Earth will be long gone as a habitable world, and the Sun will be nearing the last part of its career as a luminous star; it may even have already become a white dwarf. We will not be there to see it – but will anybody?

Much of the released energy will be dangerous, for example in the form of X-rays, and any life-bearing planets will be deluged with high-energy radiation that will disrupt metabolic processes and damage living tissue. The radiation may well be sufficient to wipe out even the most technically advanced civilizations. At least we may be confident that finally the activity will subside, and the newly formed galaxy will settle down. Most of the gas will have been used up in the fireworks that followed the collision, and so the rate of star formation too will have peaked. Perhaps the eventual outcome will be a system that is calm and stable, but also lifeless.

Throughout the next five billion years from now these various processes will continue – star deaths, star births, supernovae and collisions between galaxies. The most significant long-term change will be the increasing distances between the clusters of galaxies. We are drifting slowly but inexorably into the long twilight of the Universe.

18.7 BILLION YEARS A.B. (AFTER THE BANG) ONWARD

CHAPTER 7 **The End of the Universe**

▲ Graveyard of the Universe

In 10^{20} years time, vast black holes (seen at right) will share a massively expanded Universe with the remains of stars and planets.

▶▶ Supermassive black hole (NGC 1097)

This image shows in detail the channelling process by which matter is swallowed by the supermassive black hole at the centre of the spiral ring. More than 300 star-forming regions are visible as white spots along the ring of dust and gas.

W hat is the ultimate fate of the Universe? It is hard to choose between a range of possibilities, but the answer must depend on the relative strengths of the only two players in the 'endgame' – gravity, and the force that drives the acceleration of the universe (called the 'cosmological constant').

Let us examine a future in which gravity wins. The expansion would come to a halt, and then reverse. Instead of observing galaxies moving away from us with redshifted spectra, we would see blueshifts as they move toward us. The temperature of the Universe would rise, and collisions between clusters of galaxies would become increasingly common. The sky would brighten, and eventually the entire Universe would end in what has been called the 'Big Crunch', something like the Big Bang in reverse.

What happens then? Perhaps the Universe could rebound so that our Universe's Big Crunch becomes the Big Bang of the next one, and so on into infinity. This cosmic recycling allows us to escape from having to suppose a moment of creation in which time began – and there is something comforting about that idea.

Unfortunately, current evidence offers little chance that the Big Crunch will ever happen; there simply is too little matter in the Universe (even including dark matter) to reverse the expansion. Gravity is not strong enough. The presence of the second player, the cosmological constant, only makes things worse, and it seems the Universe will expand for ever at an ever-increasing rate. As what happens next depends on the strength of the 'cosmological constant', this is a good time to ask whether it is indeed a constant. As yet, we simply do not have the evidence to decide. Currently, everything we know is consistent with it having a constant strength, so let us assume this, and see what happens.

Expanding forever

Long after our Sun has become cold and dead, stars will shine on as the distances between the clusters of galaxies go on increasing. It is thought that over the relatively short distances between members of these clusters, gravity will remain dominant, and strong enough to keep them together. But, over the huge distances between groups, the cosmological repulsion force brings about an ever-increasing gulf. Galaxies as viewed from each other will become very dim, and even within clusters there will be changes. As time goes by, the brilliant stars will explode to leave feeble remnants, and there will be increasing numbers of black holes. With less and less matter available to form fresh stars, there will be at first a gradual and then an accelerating descent into darkness.

From perhaps 10^{13} years hence the stars will have ceased to radiate; there will be no nuclear reserves left. Gravitational effects continue to operate, and there will be many close encounters between black dwarf stars. A star moving round the centre of a galaxy will lose energy by radiating gravitational waves, and will slowly migrate toward the galactic centre; it will join others, and the result will be the formation of supermassive black holes. It may be that the same basic principles apply to the members of what used to be a supercluster of galaxies, such as that which today includes the Local Group and also the Virgo Cluster, with all the matter collecting at the centre.

After about 10^{20} years, which is ten billion times longer than the present age of the Universe, the scene will indeed be dismal; dead stars, the ghosts of planets, vast black holes, and elementary particles and photons spread out. The whole of space will have increased to a scale beyond our comprehension; the black holes will be separated by a distance at least a hundred times greater than the present extent of the observable

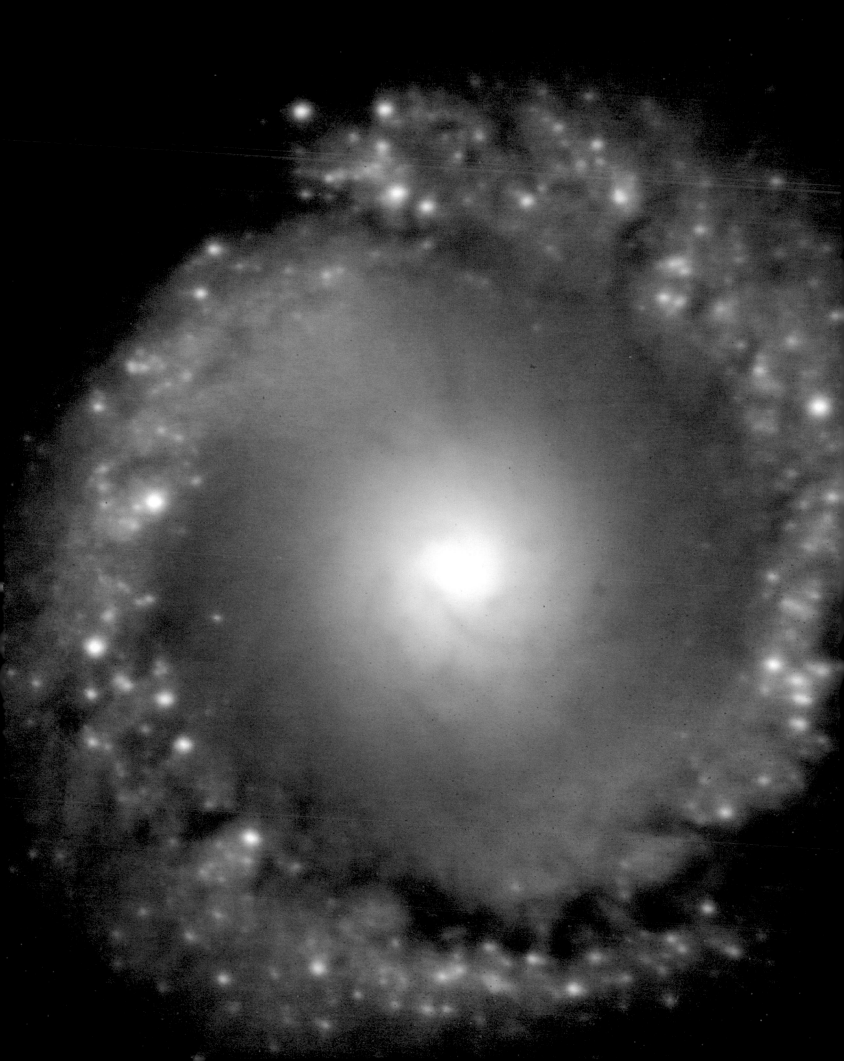

▲ The death of a black hole

This is a visualization of a black hole shrinking through emission of Hawking radiation. It is believed that ultimately all black holes will end their lives in a burst of radiation. As it shrinks, so the radiation which it emits gradually shifts toward the blue end of the spectrum.

Universe. Life everywhere will have become extinct. The Universe will not yet be dead, but the game is nearly over.

Nothing is forever

Even the black holes may not be permanent. We mentioned earlier that the vacuum of any volume of space is thought to be filled with what are termed virtual particles, which have lifetimes so short that they cannot usually turn into ordinary matter. These particles appear in pairs that are identical in every way except they carry opposite charges. The pairs quickly annihilate each other.

However, suppose a particle and its antiparticle appear just outside the event horizon of a black hole; remember that the event horizon is the boundary of the region from within which there is no escape. Before the pair can annihilate each other, as would ordinarily happen, one member of the pair may be sucked across the event horizon, while the other is ejected in the opposite direction. To an observer outside the black hole, this is tantamount to saying that the black hole has emitted a particle from within the event horizon, so that in effect the mass of the black hole decreases by the same amount as the mass of the emitted particle; the radius of the event horizon also shrinks. This can happen time and time again. The black hole becomes smaller and smaller with the emission of what is called Hawking radiation, and finally it evaporates in a final burst of radiation.

Then the ultimate: proton decay. A proton is thought to be made up of particles called quarks, but it may eventually disintegrate into lighter particles plus radiation. It may first decay into a positron (an antielectron) and a particle called a pion, which is so unstable that it would promptly decay into photons. The average lifetime of a proton is estimated to be of the order of at least 10^{31} years, so that it is hardly surprising that no instances of proton decay have been found as yet – the universe is only 10^{10} years old. But if this scenario is right, then in 10^{33} years hence there will be nothing left but a sea of photons and elementary particles.

The expansion of space will lead to an unbelievable dilution. And it has been estimated that in 10^{66} years time the average distance between typical electrons will be over a hundred thousand times the radius of the Universe we can examine today. A googol (10^{100}) years will pass; by perhaps 10^{116} years the remaining particles will decay into

radiation. The Universe has become steadily darker and cooler, and nothing more is ever likely to happen.

Remember that this description is based on the assumption that the accelerating force remains constant in strength. What if it doesn't? It seems that if it weakens – or even if it stops acting completely – we still reach, more slowly but just as inevitably, the dismal, isolated future we have just described. But if the cosmological constant were to increase in strength, a more dramatic end would await us.

The Big Rip

At first there will be little noticeable difference. Things may happen more quickly, but we will soon be left with an isolated cluster of galaxies, which until now has always been held together by gravity. Gravity is strongest when objects are close together – on small scales – whereas the accelerating force increases in strength with separation. Eventually,

▲ **Black holes – the last survivors**
Galaxy M87 is a prominent member of the Virgo cluster of galaxies, 60 million light-years away. It has 14,000 globular clusters and spectacular jets, which extend over 8000 light-years from its centre. It radiates powerfully across the spectrum from X-rays to radio. Powering all this is a massive black hole.

though, the increasing strength of the cosmological constant will dominate on smaller and smaller scales. First the clusters of galaxies will be torn apart, leaving just an isolated galaxy in the centre of the observable universe. Structure in the Universe has, at this stage, less than a billion years to go. With sixty million years to go before the end, the individual galaxies will be torn apart, sending stars – or at least the remnants that remain – flying in all directions. The Universe is now more empty, and matter more isolated, than it is possible to imagine, but this Universe has more drama to come, in the form of what has become known as the Big Rip.

As the Universe's expansion continues, getting ever faster, eventually the matter making up the stars themselves will be ripped apart – any planets still surviving will be destroyed just thirty minutes before the end. We will be left with a sea of atoms. Even this is not the end, as if the expansion continues to accelerate the atoms themselves may be pulled apart, leaving only radiation. Even the forces that hold the atomic nucleus together can no longer resist the repulsive force, and the Universe is left as a sea of radiation and particles, much as it was just after the Big Bang but almost infinitely less dense.

This is sober science – and yet one is bound to have an instinctive, rather eerie feeling that something is wrong. A Universe that ends in any of these ways seems pointless, and there may well be a vital factor that we are missing. If at last no more events can take place, then we have nothing to measure and we may as well say that time has ended. If time ends, then we cannot speculate as to what would happen after that, because there would be no 'after'. It is hard to credit that this immensely complex and ordered universe can end in featureless chaos. Science can take us no further, and unless our powers of intellect can develop sufficiently to give us a new insight, we can do no more.

Parallel universes

At least we know that our Earth, and life upon it, has a limited future. The Universe has a more extended future, but if modern theories are correct this future is not indefinite. Therefore, does this mean that there will come a time when all intelligence ceases? We know a great deal about our Universe, but there is also the concept of 'parallel universes', coexisting with ours but in a different dimension that makes contact impossible. Such a universe may differ from ours in every way; differently made up, with a different origin, and with a different timescale. If parallel universes exist, will they, too, face extinction?

Assuming they exist, and of this we do not have the slightest proof, parallel universes may endure long after ours has either died or vanished, and if they support intelligent life, then the picture of final, complete inertness may not be valid. The trouble here is that at present there seems no way of finding out. If we do come to a final state of total inactivity, then it may be said that the whole 'experiment of the universe or universes' may have been futile, and this is something that many people will find hard to accept.

The end of the story

We have done our best to trace the history of the Universe as it is currently understood by astronomers. We began with the Big Bang, and we have journeyed through the eras of inflation, transparency, first stars, galaxies, planets and life; and further, into the literally dim and distant future of billions of years hence. At the time of writing, this story is still the most convincing one we have, but will it still be accurate in a hundred years time, or even a decade? Well, we just don't know!

▶▶ **The Big Rip**
Even the forces that hold the atomic nuclei together can no longer resist the repulsive force and the Universe is left as a sea of radiation and elementary particles, much as it was just after the Big Bang, but almost infinitely less dense.

EPILOGUE

'We are stardust – we are golden…' (Joni Mitchell, 1969)

Every day, with the ever more powerful instruments at their disposal, and with penetrating theoretical work reinforced by sophisticated computer models, astronomers are learning more about the exquisitely complex Universe we live in. The more data are gathered, the more it seems that every type of event that followed the Big Bang was a necessary part of creating a blue planet that could make possible the evolution of the the human race, and all the other species of animals and plants that have been born alongside us. This does not mean that we are any more important, of course, than any of our fellow creatures, or that our time is more significant than any other chapter in the unfolding of the story of the Universe. Perhaps the evolution of Mankind is merely a necessary part of the way the Universe will evolve in the future.

We hope that, in telling this story continuously in its correct order, rather than in the order in which the discoveries were made, we have communicated to the reader a feeling of the extraordinary power of the thread that runs through the evolution of the Universe – a thread into which we are all inextricably tied. Everything that we are, and everything we know, was there in that first Big Bang. And in a sense we are still in it.

We, the authors, believe that this history book is a reasonable attempt to portray *how* things happened, according to current knowledge. Very deliberately, we have not ventured one inch into the territory of *why* they happened, or into the question of whether some extra-human intelligence put the whole sequence into operation. This, in the absence of any physical evidence, is the realm of mysticism and religion. We feel that if the workings of the Universe, in all their beauty, are properly appreciated, there is no conflict between Science and Religion; they merely deal with different areas. We hope that in a small way we will have contributed to calmness and lightness in the discussions that currently rage around our planet on this subject. If we spent a lifetime trying to understand completely how a single daffodil is made, we would be no nearer to understanding why such beauty is shown to us; nevertheless, we can have endless fun satisfying our curiosity in both areas. We wish you endless fun.

Brian May, Patrick Moore and Chris Lintott, 2006

◄◄ **Our home in space**
Just another picture of the crescent Moon? Look again…. This is our own blue-green planet seen from space. All human history happened here.

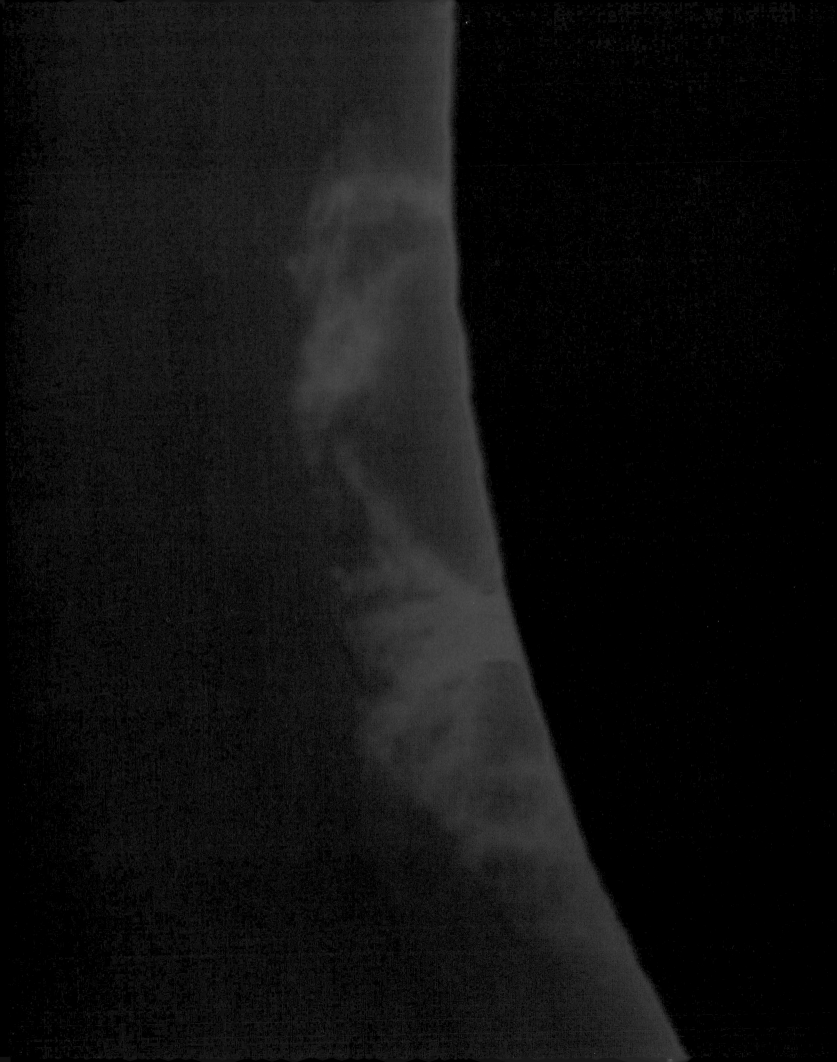

PRACTICAL ASTRONOMY

How much of this amazing Universe can you see personally? The answer is: a great deal. Astronomy is just about the only science where amateurs and professionals work closely together, and where amateurs can make really valuable contributions, as well as giving themselves endless enjoyment. And to become a practical observer, you need not spend large sums of money upon expensive equipment. George Alcock, the most famous comet-hunter and nova-hunter, never used a telescope in his life. All his work was carried out with powerful, specially-mounted binoculars, which he took out into his garden!

How to become an astronomer

The following suggestions may be useful, though of course different people will have different ways of doing things, and a great deal depends upon personal circumstances and environment. But we give them as what we hope are useful tips.

1. Read some introductory books, and make sure that you know the main essentials. Having read through the previous pages you will certainly know the facts already, but in case of possible confusion, our glossary will help.

2. Join an astronomical society. There are national societies, such as the British Astronomical Association or the American Association of Variable Star Observers – all you need is enthusiasm. Moreover, most large towns and cities all over the world have local societies. By joining, you will make many new friends, and there are always people to give help and advice. In general, astronomers are sociable folk!

◄◄ **Solar prominences**

An image of a solar prominence made using the red light produced by hot hydrogen gas above the surface of the Sun.

◄ **Observing the transit of Venus, 2004**

This rare event was enjoyed by a large party of astronomers at Patrick's observatory in West Sussex, England.

3. Obtain an outline star map like those on pages 164–8, go outdoors when the sky is dark and clear, and learn your way around the constellations. This is not nearly as difficult as might be thought. Remember, you can never see more than around 3000 stars at any one time without using optics, and the patterns are distinctive. Moreover, the stars stay in the same positions for year after year, century after century. It is only our nearest neighbours, the members of the Solar System, that wander about from one constellation to another, and even they keep strictly to the belt round the sky known as the Zodiac.

4. Obtain a pair of binoculars, and start seeking out selected objects such as red stars, clusters, nebulae or the phases of Venus, and the satellites of Jupiter. You may, of course, be lucky enough to be greeted by a comet. In 1997, for example, the magnificent comet Hale-Bopp hung in the sky for months.

5. By this time you will almost certainly have decided how deep your interest really is, and what branch of astronomy attracts you most. Upon this depends what equipment you will need. You will almost certainly require a telescope, and the situation here is much better than it used to be only a few years ago, because it is possible to buy a reasonably useful telescope for less than £100 or US$200. Of course, a working observatory equipped with a large telescope and auxiliary equipment costs a great deal of money, but this can come later. Also, remember that a telescope does not wear out; provided it is properly cared for, it will last for a lifetime and more with only straightforward maintenance.

We will say more about telescopes later. Meantime, let us put ourselves in the position of enthusiasts who have read this book and are now anxious to see some of the objects about which they have been reading. Also, let us assume that they are the proud owners of good binoculars plus an inexpensive telescope – say a 3-inch or 80-millimetre refractor. What is the starting point?

The Sun

Learning the constellations is important, but perhaps most people will feel like starting with our nearest neighbours, the members of the Solar System. First on the list is the Sun – but bear one thing in mind from the outset: **the Sun is dangerous**. Staring straight at it, even when it is low down and looks quite harmless, is emphatically not to be recommended, because it emits radiation all over the electromagnetic spectrum, not only in the visible range, and obviously it is hot. To look directly at it with any telescope, or even binoculars, will cause permanent eye damage and possible permanent blindness unless suitable precautions are taken. Unfortunately, some small telescopes are still sold with 'sun caps', which it is said can be fixed in front of the telescope eyepiece to cut out the brilliance and the heat. These devices are ineffective and highly dangerous. They do not give full protection, and they are always liable to shatter without warning. They should never be used. There are filters that can be fitted over a refracting telescope, but do not try direct solar observation of this kind until you really know just what you are doing.

The answer is to use the telescope as a projector, and look at the Sun's image on a suitably positioned screen (see page 105). This is safe enough, and the image will show any sunspots that happen to be around. Observe daily, and you will see the spots being carried across the disk because of the Sun's rotation. This is fascinating, and remember that the Sun is a normal star, the only one close enough to be studied in real detail.

▼ **Sunspots**

Photographed by John Fletcher from Patrick's observatory.

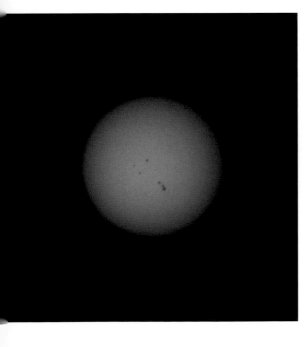

◀ **Lunar crater**

The 60-mile (96 km) diameter crater at the bottom of
the picture is named Plato.

Serious solar research involves specialized equipment beyond our present scope, but
there are plenty of books available. Certainly the solar observer need not brave the chill
of dark nights, and even from central city locations the Sun can sometimes be glimpsed!

The Moon

Come next to the Moon, which is 'safe' although it may dazzle you – the amount of heat
it sends us is too small to be a hazard. The main features are the seas (or 'maria'), the
mountains and the craters. The naked eye shows the main maria, and binoculars will
reveal many of the craters as well as the impressive mountain ranges. With a telescope,
an incredible amount of detail is available.

The maria are lava plains, some regular in outline and others less well defined. They
were once thought to be real seas, or at least sea-beds, and have retained their interesting
names: Mare Serenitatis (Sea of Serenity), Sinus Iridum (Bay of Rainbows), Oceanus
Procellarum (Ocean of Storms) and so on, but we now know that there have never been
areas of open water on the Moon. The lunar seas are bone dry, and are scarred with
craters; some of them have mountainous borders – thus the regular Mare Imbrium is
bounded in part by the Apennines and the Alps, with peaks higher than any in the Earth
ranges with the same names. Note that most of the maria form a connected system, the
main exception being the relatively small Mare Crisium (Sea of Crises). The maria do not
extend over the Moon's limb, and this leads on to a very important point.

The Moon's orbital period is 27.3 days. As we have seen, its axial rotation period is
exactly the same, so that the Moon's rotation is 'captured' or 'synchronous'. There is no
mystery about this – tidal friction over the millennia has been responsible, and all major
planetary satellites behave in the same way. The Moon keeps the same hemisphere turned
Earthward all the time, and before the Space Age we had no direct information about the

▲ Lunar eclipse

The colour of the Moon during a total eclipse depends on the state of the Earth's atmosphere. After a major volcanic eruption, like Krakatoa in 1883, the Moon looks very dark indeed.

averted regions, which are always turned away from us and which we can therefore never see. Note that, as we have said earlier, although the Moon keeps the same face turned toward the Earth, it does not keep the same face permanently turned toward the Sun, so that day and night conditions are the same for both hemispheres—there is no 'dark side'.

However, again as mentioned earlier, things are complicated by effects known as librations. The Moon's orbit is not circular, but markedly elliptical; following Kepler's laws the Moon moves quickest when closest to us (perigee) and slowest when furthest out (apogee). Yet its rate of axial rotation remains constant. This means that during each circuit, its position in orbit and the amount of rotation become slightly out of step. The Moon seems to oscillate very slowly and very slightly; we can peer a little way round, first one mean limb, then round the other. This is called libration in longitude. It, together with less important librations, means that all in all we can examine 59 per cent of the total surface, though of course no more than 50 per cent at any one time.

The remaining 41 per cent was unexplored until 1959, when the Russians sent their uncrewed probe Lunik 3 on a round trip and obtained the first pictures of the hitherto unknown regions. Not surprisingly, these areas turned out to be very like the areas we have already known, with mountains, valleys and craters. One vast sea, the Mare Orientale (Eastern Sea) lies almost wholly on the far side, though a tiny portion can be observed from Earth at maximum libration, and was first noted in 1948 by a certain astronomer named Patrick Moore! American observers rediscovered it around ten years later.

A crater is at its most prominent when at or near the terminator (the boundary between the sunlit and dark hemispheres), as its floor will be wholly or partly filled with shadow. Many large craters have high central peaks, which catch the Sun's light while the main floor is still dark. As the Sun rises, the shadows shrink, and even a large crater may become difficult to identify under high illumination unless it has an exceptionally dark floor or exceptionally bright walls.

For the beginner, full Moon is the very worst time to begin observing; there are virtually no shadows, and the scene is dominated by bright rays issuing from a few craters, such as Copernicus in the Oceanus Procellarum and Tycho in the southern uplands. The rays are surface deposits, not seen under low-light conditions. In case you are wondering about the names of the craters, they honour personalities including past observers of the Moon. The system was due originally to the Jesuit astronomer Riccioli, who drew a lunar map in 1651, and has been extended since, though some unusual people have found their way there; Julius Caesar has a large crater, not for his military prowess but because of his association with calendar reform. One crater is called Hell, but is not particularly deep; it honours the 18th-century Hungarian astronomer, Maximilian Hell!

Other lunar features include ridges, isolated peaks, low mounds or domes, and the crack-like rills (also known as rilles or clefts). Because the scene changes so quickly with the conditions of illumination, it is wise to take several formations and sketch them under different lighting, using an outline map. Persevere and it will not take you long to find your way around the surface of the Moon. When beginning an observing session, it is sensible to start out by scanning the area to be studied; by all means use a photographic outline (taking full account of the main features). Do not attempt to draw too large an area at any one time, and do not use too high a power; if the image becomes hazy or unsteady, change at once to a lower magnification. Electronic equipment used with a modest telescope can produce superb pictures.

Eclipses

When the Moon passes into the cone of shadow cast by the Earth its supply of direct sunlight is cut off, and the Moon turns a dim, often coppery colour until it passes out of the shadows again. It does not vanish completely, because some rays of sunlight are bent on to it by way of the shell of atmosphere surrounding the Earth. Lunar eclipses may be either total or partial – and obviously they can only happen at full Moon! They are not important, but they are lovely to watch and excellent to photograph.

The planets

The planets have always been favourite targets for owners of small or medium-sized telescopes, and until less than a century ago it is probably true to say that most of our knowledge of their surface detail was due to amateurs. This is no longer true, but planetary observations are as fascinating as ever, partly because one never knows what will happen next.

The inner members of the Sun's family, Mercury and Venus, are not wildly exciting. Mercury is not likely to be seen at all except if a deliberate search is being made, because it always stays close to the Sun in the sky; with the naked eye it can only be seen when at its best, either low in the west after sunset or low in the east before dawn. The best views are obtained during daylight, when Mercury is high up – but so is the Sun. To locate Mercury you need a motorized telescope with accurate positioning equipment. Sweeping around for the planet is most unwise; sooner or later the hapless observer will look at the Sun by mistake. Even when Mercury is found, all that will be seen will be the characteristic phase. Only space vehicles have so far shown the Mercurian craters, plains and mountains.

Telescopically, Venus is only slightly more rewarding. It is much brighter than any other star or planet, and really keen-sighted people can see it with the naked eye even in broad daylight. The phase is very evident, but generally the disk will appear almost featureless, with no markings apart from vague cloudy patches and bright areas. The dense atmosphere hides the surface; ordinary telescopes will not penetrate it; there is no such thing as a sunny day on Venus. Space research methods have had to be used to reveal the craters, the lava-flows and the volcanoes, and we now know that although named after the goddess of love, Venus is a hostile place. There is little scope for the visual observer.

Transits

There are occasions when Mercury and Venus pass in transit across the face of the Sun. Mercury does so reasonably often – the next occasion will be May 9, 2016 but will only be observable with optical aid. The next transit of Venus will be on June 6, 2012, after which there will be no more until December 11, 2117. Venus is quite conspicuous in transit – but of course, when observing transits all the normal precautions for solar observing must be taken.

Mars

Mars is different because the atmosphere does not hide the surface features, and when the planet is well placed, a small telescope will show the dark areas, the ochre 'deserts' and the white polar caps. Yet Mars is a small world, and is seen well for only a few weeks when the planet is in opposition – positioned on the opposite side of the Earth

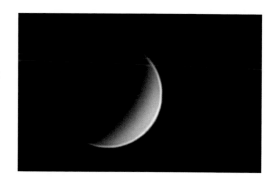

▲ The crescent of Venus

Some observers have reported an ashen light on the dark side, which has been attributed to everything from lightning to city lights under the clouds! The most likely explanation, perhaps sadly, is an optical illusion.

▼ Transit of Venus

Brian took this picture of his projection of the 2004 transit of Venus on a card at Patrick Moore's observatory in Sussex.

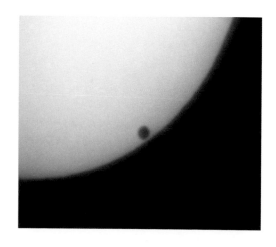

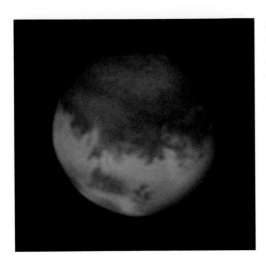

▲ **Mars**

The Martian polar caps wax and wane with the seasons and are cleary visible in small telescopes. The dark areas are sometimes obscured by dust storms, which start small but can engulf the entire planet.

from the Sun. Moreover, it does not bear magnification well, so that the observer has to take advantage of the best nights around the time of opposition. Not all oppositions are equally favourable; the orbit of Mars is markedly eccentric, and the best oppositions occur when the planet is at opposition and perihelion (closest point to the Sun) at the same time. In 2003, the distance to Mars was less than 35 million miles.

Away from opposition, Mars shows a distinct phase, resembling the Moon a few nights from full. When making drawings, always remember to allow for phase. The rotation period is over half an hour longer than ours, so that there is no urgent rush to position the main features, as there is with Jupiter.

If you are making a sketch, first, put in the polar cap (if visible) and the dark markings. Look round for any cloud phenomena, change to the highest available power, and add the fine detail. Record the time of observations plus the details of the telescope, magnification and weather conditions and also the longitude of the central meridian. Be careful to draw only what you can see – not what you expect to see. Patrick has vivid memories of his first view of Mars through the 24-inch refractor at the Flagstaff Observatory in Arizona, used by Percival Lowell to draw his famous canal network. Would canals be seen? It was a relief to find that they were conspicuous only by their absence.

Visual observations of Mars are being to some extent supplemented by the excellent photographs obtained by amateurs using electronic instruments in combination with small telescopes. The images shown in this chapter are better than any professional photographs available of Mars in, say, 1960. We can see the great volcanoes such as Olympus Mons – but from Earth we could never feast our eyes upon the complex summit caldera; neither could we ever see into the depths of the colossal Valles Marineris. Only spacecraft can show us these wonders.

Asteroids

Beyond the orbit of Mars we come to the Asteroid Belt. These midgets look like stars through a telescope, after all even the largest of them, Ceres, is a mere 600 miles in diameter. However, detailed times of their positions are available, and asteroids are easy to photograph. Some asteroids, such as Hermes, are classed as PHAs (Potentially Hazardous Asteroids), because their orbits cross that of the Earth. Future collisions cannot be ruled out and amateur astronomers have a role to play here. There so many PHAs that professional astronomers are hard pressed to keep track of them all.

Jupiter

Jupiter and Saturn, the giants in the Sun's family, are of special interest to the amateur observer. Jupiter, with its belts, its spots and its Galilean satellites, is always changing. Saturn's rings make it arguably the most beautiful object in the sky.

Jupiter, large enough to hold over a thousand globes the volume of the Earth, shows a yellowish, flattened disk crossed by cloud belts. Generally there are two main belts, one to either side of the Jovian equator, with others at higher latitudes. Jupiter has a gaseous surface, and does not rotate in the same way as a solid body would do. In what is termed System I (between the north edge of the South Equatorial Belt and the south edge of the North Equatorial Belt) the rotation period is 9 hours 51 minutes. Over the rest of the planet (System II) the period is five minutes longer, and individual features, such as spots, have periods of their own, so they drift around in longitude.

The belts, due to droplets of liquid ammonia, dominate the scene. Spots can become more prominent but usually only for a limited period. The exception is the Great Red Spot, which has been seen ever since the 17th century; it is brick-red, and over 20,000 miles (30,000 km) long. It can vanish for a while but always returns. There are some well-marked white spots, but these are always temporary; in 2006 a smaller red spot was seen.

Jupiter is a quick spinner, so that one cannot afford to linger when making a sketch; the main drawing should be finished in less than 10 minutes, with finer details added afterwards as quickly as possible. Rotation periods of discrete features can be found observationally. Time the moment when the feature crosses the central meridian, and then use tables to work out the longitude. This is not difficult – the central meridian is easy to find because the planet's disk is so flattened. With practice, estimates can be made to an accuracy of within a minute (sixtieth of a degree). Work of this kind used to be very valuable. We have to admit that this is no longer so because of the improvements in imaging. But Jupiter remains fascinating, and quite apart from the surface details watch too for the transits, occultations and eclipses of the Galilean satellites.

For sketches, it is sensible to use prepared blanks. Make sure to look for anything unusual. In 1994 the fragments of Comet Shoemaker-Levy 9 cascaded into Jupiter, causing scars that persisted for months. Any small telescope was capable of showing them when they were first formed.

Saturn and beyond

Saturn's rings are able to provide endless enjoyment. It is interesting to compare the amateur photograph overleaf taken with an 15–inch telescope with a view from the Hubble Space Telescope. You have to look carefully to decide which is which. Occasionally white spots appear on the disk. One was discovered in 1933 by Will Hay, the stage and screen comedian. Another spot of the same kind was discovered in 1990 by an American amateur, Stuart Wilber. Saturn is a less active world than Jupiter but it is still capable of springing surprises.

Photographing the outer giants, Uranus and Neptune, is very easy, though of course no surface details can be seen. There are also the bodies of the Kuiper Belt, of which Pluto is the brightest though not the largest. Imaging these objects is really useful in helping to keep track of them.

▶ **Saturn's rings**

Saturn's beautiful rings are not always visible in small telescopes. This view was created from a stack of 9000 separate images.

Finally, in the Solar System, we come to the most erratic members of the family; comets and meteors. There are always various comets within range of modest telescopes, though really spectacular visitors, such as Hyakutake and Hale-Bopp are depressingly rare. Some comets look rather like tiny patches of shining cotton-wool, and to locate them you will need a detailed star atlas and a telescope fitted with accurate setting circles. Of course really spectacular comets, such as Hale-Bopp, are well seen in binoculars.

▶ **Comet Hale-Bopp**

Photographed by Patrick in 1997, this lovely comet will not return for 2360 years. Note the proximity to the chimney stack of Patrick's house.

Meteors can be seen at any time but the annual showers are interesting to watch; good photographs can be taken with an ordinary camera. The August Perseids are always reliable; given a clear dark sky for a few minutes between around August 9 and 18 you will be very unlucky not to see several Perseids.

The stars of the northern hemisphere

Come now to the stars. This is no place to give a full review of the constellations so we have given just a few charts, which will, we hope, enable the newcomer find his way around. For northern-hemisphere observers there are three key groups, Ursa Major, Orion and Pegasus, so let us start there.

Ursa Major, the Great Bear, is circumpolar from the latitudes of London or New York – that is to say, it never sets, so that it can always be seen somewhere or other provided

▼ Zodiacal light

The evening zodiacal light in Gemini photographed in Tenerife in 1971. It was taken by Brian, with a Pentax 35mm film camera on a tripod. As the Earth rotated during a time exposure of a few minutes, the stars left trails of light on the image. The zodiacal light is a cone of light visible in the evening or the morning. It is believed to be sunlight reflected from dust in orbit around the Sun. Brian's PhD research was concerned with determining the motions of the dust.

that the sky is sufficiently dark and clear. Seven of its stars make up the pattern known familiarly as the Plough or (in North America) the Big Dipper. Because they are so useful, we have named these stars in Map 1: Alkaid, Mizar, Alioth, Megrez, Phad, Merak and Dubhe; their magnitudes are between 1.7 and 2.4 except for Megrez, which is considerably fainter. Merak and Dubhe are known as the Pointers, because they show the way to Polaris, the Pole Star, in Ursa Minor (the Little Bear). Polaris is within one degree of the north celestial pole, and so seems to remain almost stationary in the sky. Ursa Minor contains only one other reasonably bright star, the orange Kocab (2.1).

On the far side of Polaris with respect to Ursa Major is Cassiopeia, whose five main stars form a *W* pattern (magnitudes 2 to 3). Like Ursa Major, Cassiopeia is circumpolar.

▶ **Map 1**
Ursa Major.

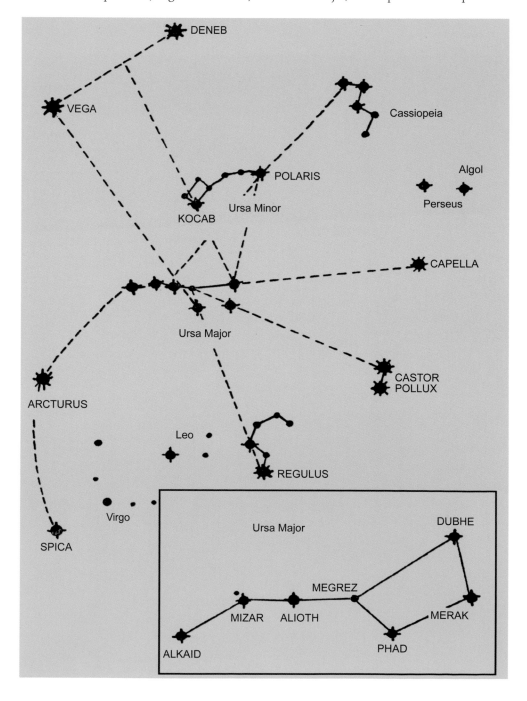

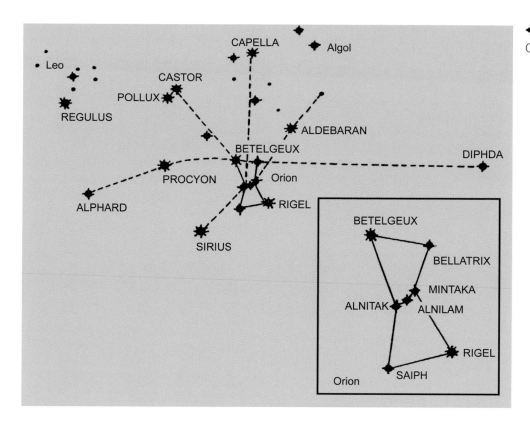

◄ **Map 2**

Orion.

When the Bear is high in the sky Cassiopeia is low down, and vice versa. From the Bear you can find the brilliant bluish-white Vega (0.1) in Lyra (the Lyre), and also Deneb (1.3) in Cygnus (the Swan). Both these are circumpolar from latitudes of Britain and the northern United States, though at their lowest they almost graze the horizon.

The remaining stars in Map 1 are not circumpolar. The 'tail' of the Bear shows the way to the orange Arcturus (-0.06) in Bootes (the Herdsman). And then to Spica (1.1) in Virgo (the Virgin). From Megrez and Merak you can locate Castor (1.5) and Pollux (1.2), the Twins (Gemini); Pollux, the brighter of the two, is orange, while Castor – a fine binary – is white. Also find Leo (the Lion), led by Regulus (1.4); extending from Regulus is a curved line of stars known as the Sickle.

Map 2 shows Orion, the celestial hunter, with his brilliant retinue. Orion dominates the evening sky in northern winter (the southern summer), and cannot possibly be mistaken. The leading stars are Rigel (0.2) and Betelgeux (variable between 0.3 and 0.8); Rigel is glittering white, while Betelgeux is the orange-red supergiant. The other main stars of Orion are around the second magnitude; Alnitak, Alnilam and Mintaka make up the belt – immediately south of that, visible with the naked eye as a misty blur, is the sword, containing the famous Orion Nebula. Southward, the belt shows the way to Sirius (-1.4) in Canis Major (the Great Dog), which is pure white, but when low down seems to flash all the colours of the rainbow! Northward, the belt points to the orange Aldebaran (0.9) in Taurus (the Bull) and on to the magnificent open cluster of the Pleiades, or Seven Sisters. Aldebaran, the 'Eye of the Bull' looks much the same colour as Betelgeux, but is not nearly so luminous – minus 140 times as powerful as the Sun. Extending from it in a *V*-pattern are the stars of the Hyades cluster, but Aldebaran is not a cluster member; it is only 65 light-years away, and just happens to lie between the Hyades and ourselves.

▶ **Map 3**

The Summer Triangle and adjacent
constellations.

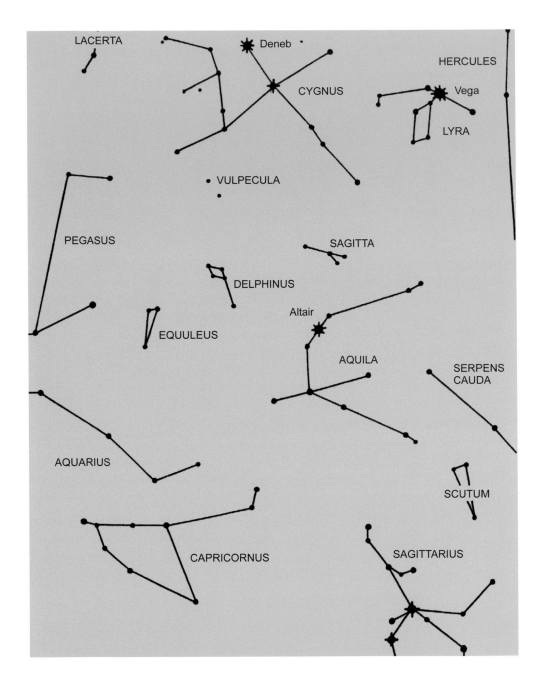

Map 3 shows what is known commonly as the Summer Triangle (an unofficial name,
popularized by Patrick during a television broadcast around 1960!). It consists of Vega in
Lyra, Deneb in Cygnus and Altair (0.8) in Aquila (the Eagle), and is linked to either side
by fainter stars. Lower down is Sagittarius (the Archer) with the superb
star-clouds in the direction of the centre of the Galaxy.

Map 4 shows Pegasus, whose four main stars make up a square; Markab (2.5), Algenib
(2.9), Scheat (variable from 2.4 to 2.9) and Alpheratz (2.1). For some unknown reason
Alpheratz has been given a free transfer to the adjacent constellation of Andromeda, with
Mirach and Almaak (each 0.1). Here we see the great spiral galaxy M31. It is just visible
with the naked eye; binoculars show it easily, though only as a misty patch. The map also
shows Fomalhaut (1.2) in Piscis Australis (the Southern Fish), which is always very low

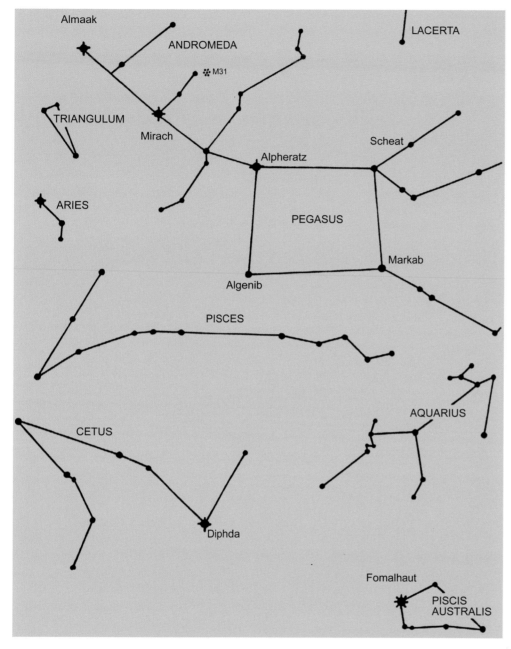

◀ **Map 4**
Pegasus and adjacent constellations.

down from northern latitudes. Pisces (the Fishes), and Aquarius (the Water-bearer) are dim Zodiacal constellations; Aries (the Ram) does have one bright star, Hamal (2.0).

Sky survey

Next let us survey the sky, soon after darkness falls, for each season: **Winter**: Let us say mid-January at 10 pm (22.00 GMT). Ursa Major is in the southeast, Cassiopeia is in the southwest; Capella is almost overhead, with Vega very low in the north. Pegasus is setting in the west, Orion is high in the south, with Sirius outstanding and Leo rising in the east.

Spring: Mid-April, 22.00 GMT. Ursa Major is overhead; Capella is descending in the west, with Vega gaining altitude in the east. Cassiopeia is in the north. Orion has almost

▶ **Map 5**

The major constellations of the
southern hemisphere.

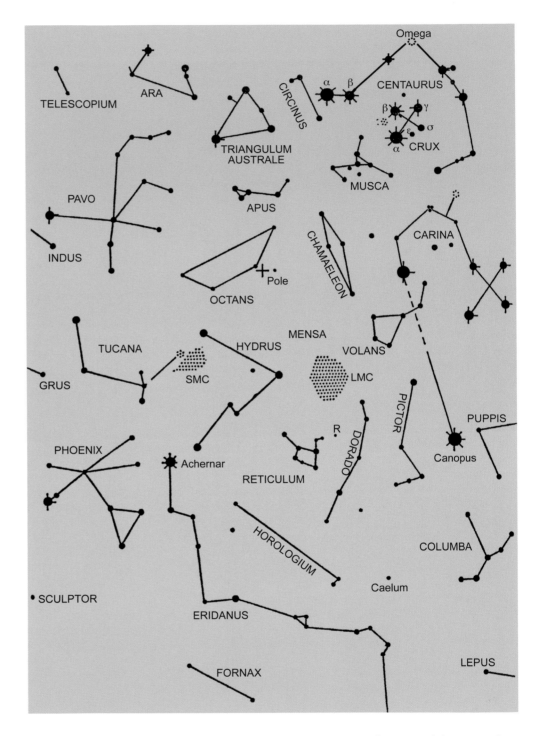

disappeared, but Sirius is still visible; Arcturus is prominent in the east, while Leo and
Virgo are high in the south.

Summer: Mid-July, 22.00 GMT. Ursa Major is in the north-west, Cassiopeia is in
the north-east; Vega is almost overhead, and Capella is near the northern horizon. The
Summer Triangle dominates the south. Arcturus is fairly high in the west, Virgo is setting;
Sagittarius is low in the south. Much of the southern aspect is occupied by large, dim
constellations – Hercules, Ophiuchus and Serpens, with only one bright star, Rasalhague
(2.0) in Ophiuchus, the Serpent-bearer.

Autumn: Mid-October at 22.00 GMT. Ursa Major at its lowest in the north, Cassiopeia is overhead. Pegasus is high in the south; Fomalhaut is low in the south; the Summer Triangle is still very prominent. Aldebaran and the Pleiades have risen; Orion will soon follow. Winter, with its frosts and snows, lies ahead!

The stars of the southern hemisphere

So much for the northern hemisphere. Now let us go south – for example, to South Africa or Australia. Remember that you do not actually have to cross the equator to see the Southern Cross; it rises from anywhere south of latitude 35 degrees north.

The Southern Cross, Crux, is the most famous of the southern constellations, shown in Map 5. Its four main stars are Acrux or Alpha Crucis (0.8), Beta Crucis (1.3), Gamma Crucis (1.6) and Delta Crucis (3.1). In fact Crux is shaped more like a kite than a cross; it is not nearly so cruciform as Cygnus in the northern sky. It is also much brighter and smaller – rather surprisingly, it is the smallest of all the 88 accepted constellations. It is almost surrounded by Centaurus (the Centaur); Alpha Centauri (-0.3) and Agena or Beta Centauri (0.6) point to it. The two Southern Pointers are not associated; Alpha, at 4.4 light-years, is the nearest of the bright stars, while Agena is a giant 530 light-years from us and 13,000 times as luminous as the Sun. In the Centaur, look for the globular cluster Omega Centauri, an easy naked-eye object.

Unfortunately, the south celestial pole is not marked by any bright star; we have to make do with Sigma Octantis, below the fifth magnitude and hidden by any mist or haze in our atmosphere. The Pole lies between Crux and Achernar (0.5) in Eridanus (the River). Also on this map is Canopus (-0.7), the brightest star in the sky apart from Sirius; it is a supergiant. It lies south of Orion; it can just be seen from Alexandria in Egypt, but not from the other great centre of ancient astronomy, Athens.

Next, a survey of the southern sky through the seasons. **Summer:** Observing in January in mid-evening. Orion is high in the northern part of the sky; Sirius is very visible, Canopus is almost overhead, Crux is in the south-east with the Southern Pointers. Capella is very low in the north.

Autumn: Mid-evening in April. Crux and Centaurus are very high, Canopus is sinking in the west, Sagittarius is gaining altitude in the south-east. Arcturus is in the north-east, but Achernar is low in the south.

Winter: Mid-evening in July. Scorpius and Sagittarius are almost overhead and the star clusters are glorious. Crux and Centaurus are sinking in the south-west, Achenar is rising in the south-east. Vega can be seen rather low in the north; Fomalhaut is conspicuous in the south-east.

Spring: Mid-evening in September. Crux and Canopus are low in the south; Pegasus can be seen in the north and the northern Summer Triangle, here perhaps better referred to as the Winter Triangle. Fomalhaut is almost overhead.

Finally, on no account fail to look at the two Magellanic Clouds, in the south polar area. With the naked eye they look rather like detached parts of the Milky Way, but they are in fact satellite galaxies, over 169,000 light-years away. They contain objects of all kinds – giant and dwarf stars, clusters, nebulae and novae. There have been two observed supernovae, one in 1885 and the other in 1987. Northern observers always regret that they lie in the deep south. We appreciate that this ramble round the constellations is very fragmentary, but it may act as a start. The skies are never dull.

▶▶ **La Palma**

A panoramic view of the Roque de los Muchachos observatory in La Palma, Canary Islands.

BIOGRAPHIES

Wilhelm Heinrich Walter Baade 1893–1960

Walter Baade was born at Schröttinghausen in Germany; he studied at Munster and then at Göttingen University, where he graduated in 1919. After a spell in Hamburg he went to the United States in 1931, joining the staff of Mount Wilson Observatory near Pasedena, California. Though interested mainly in astrophysics, Baade also paid some attention to the Solar System, and discovered ten asteroids – among them Icarus, the first asteroid known to have an orbit carrying it closer to the Sun than Mercury.

During the Second World War, North America was under a blackout, and the nights at Mount Wilson were exceptionally dark. Baade made good use of them, and with the 100-inch Hooker reflector set out to study individual stars in the centre of the Andromeda Galaxy. To his surprise, he found that the brightest stars in the nucleus of the spiral were not bluish-white, as expected, but old red giants. From this, Baade concluded that there are definite types of what he called 'stellar populations'. Population I is made up chiefly of young, hot stars, while in Population II the most luminous stars are red giants. Spiral arms are comprised of mainly Population I stars, while the centres of galaxies are mainly Population II.

At the end of the war, Baade transferred to Palomar, and used the then-new Hale reflector to study faint variable stars. Finally, he found that there are two types of Cepheid variable; those of Population I are twice as luminous as those of Population II. To measure the distances of galaxies, Hubble and Humason had used 'the wrong type of Cepheids', so that the galaxies were twice as far away as had been believed; the Andromeda Spiral is 2.9 million light-years away, not 750,000 light-years. There had been an error of more than 100 per cent.

One of Baade's colleagues was Fritz Zwicky, but the personal relationship was not exactly happy; Baade was officially an enemy alien (though nobody in authority seemed to mind), and Zwicky referred to him as a Nazi, even threatening to kill him if he ventured alone onto the university campus. Baade took this threat seriously, particularly as Zwicky's appearance was often described as 'menacing'.

Baade stayed at Palomar until 1958, but then returned to Germany to become a professor at Göttingen University. He died two years later. He was a pleasant, friendly person, and will always be remembered as being the man who, in one short and elegant research paper, calmly doubled the size of the Universe.

Subramanyan Chandrasekhar 1910–1995

Chandrasekhar – always referred to simply as 'Chandra' – was probably India's greatest astrophysicist. His name is immortalized in connection with the 'Chandrasekhar Limit', the greatest mass that can be obtained by a white dwarf star – more massive bodies form neutron stars or black holes at the end of their life. Popular legend suggests that he completed the work for this, his greatest discovery, while on the sea journey from India to England to become a student in Cambridge following his first degree in Madras. Perhaps more astrophysicists should be sent on long sea voyages…

His initial theory was strongly criticized by some leading astronomers, who ridiculed the idea that a star could eventually pass beyond the white dwarf stage into an object

▲ **Walter Baade**

◀◀ **Large spiral galaxy NGC 1232**
Glorious images of remote galaxies like this one captured by the European Southern Observatory's Very Large Telescope are now commonplace, but results like this would not have been possible without the foundations laid down by generations of talented astronomers.

▶ Chandra's X-ray Universe

Named for Chandrasekhar, NASA's X-ray satellite
Chandra captured this image of the colliding
Antennae Galaxies.

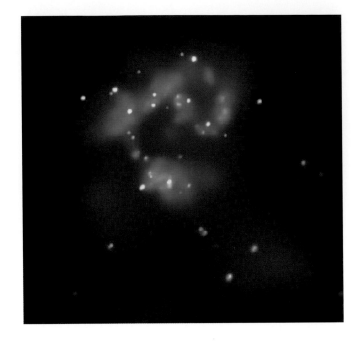

of such enormous density that, in his own words, 'one is left speculating on other possibilities'. His main adversary was Sir Arthur Eddington, generally regarded as the leading astrophysicist of the day. The arguments between the eminent professor and the young student became decidedly heated, and have gone down in history. In a letter home Chandrasekhar wrote: 'The differences are of a political nature. Prejudices! Prejudices! Eddington is simply stuck up. Take this piece of insolence: "If worst comes to worst we can believe your theory. You see I am looking at it from the point of view not of the stars but of Nature"… "Nature" simply means "Eddington personified as an angel". What arguments could anyone muster against such brazen presumption.' Amazingly, despite these quarrels, Chandrasekhar and Eddington managed to remain on friendly terms.

Chandra became Professor Emeritus in Chicago, and once drove 200 miles on a round trip to teach his class of students – to find that because of a heavy snowstorm, only two had turned up – Tsung Dao Lee and Chen Ning Yang, both of whom were subsequently Nobel laureates in 1957 – even before Chandra himself.

On July 23, 1999, NASA launched an X-ray satellite, then known as AXAF (Advanced X-Ray Astrophysics Facility), but subsequently renamed Chandra, which also means 'luminous' in Sanskrit. It was an immediate success, and was still functioning in 2004.

Arthur Stanley Eddington 1882–1944

Arthur Eddington was one of the most important astrophysicists of the 20th century. His brilliance showed at an early age; after graduation he was principal assistant at the Royal Greenwich Observatory (1906–1913) and then became a professor at Cambridge. In 1919 Eddington took part in a famous experiment. Albert Einstein had predicted that if starlight passed close to a massive body such as the Sun it would be displaced by an angle twice as large as that predicted by Newtonian theory. Only during a total solar eclipse can the stars and the Sun be seen at the same time from Earth, and in 1919 Eddington went to observe the stars during an eclipse visible from Príncipe,

▼ Eddington's eclipse negative

Taken from the report of the 1919 eclipse expedition
to verify Einstein's prediction that light would be bent
by our Sun.

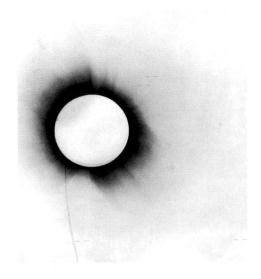

off the western coast of Africa. It is said that one of the expedition leaders asked the Astronomer Royal what would happen if the expected displacements did not show up. 'In that case,' was the reply, 'Eddington will go mad, and you will have to come back alone.'

The displacements were observed despite fairly nasty conditions, just as Einstein had predicted, though admittedly there may have been a certain element of luck in the observations. Eddington returned in triumph to Britain, announcing his results in rhyme: 'O, leave the wise our measures to collate, Light rays – when near the Sun – do not go straight.'

Incidentally Eddington, as a Quaker and pacifist, refused to take part in any military service during the First World War, but was given exemption because of his scientific achievements. The impact of an English scientist confirming a prediction made by a German in 1919 was immense, and from that moment on Einstein (and Eddington) had superstar status. The scientific community was less impressed with the cloud-affected results, and real confirmation had to wait until the next eclipse expedition in 1922.

The value of Eddington's contributions to stellar astronomy can hardly be over-estimated. His 1926 book *The Internal Constitution of Stars* is still regarded as a classic. He was also one of the main contributors to relativity theory. Apparently it was suggested to him that only three people, including Einstein himself, really had a full knowledge of relativity. Eddington paused, and thought for a moment. 'Interesting. Who is the third?'

Eddington was a brilliant popular writer, and an early radio broadcaster on astronomy. He could make mistakes, and some of his criticisms of his colleagues were – to put it mildly – robust. But he will always be remembered as one of the principal founders of modern theoretical astrophysics.

Albert Einstein 1879–1955

Few people will doubt that Albert Einstein is the greatest scientist who has lived since the time of Newton – yet his early career was anything but promising! He began as a student in Munich, and over the next few years took various academic exams, failing most of them. By 1901 he had taken a post as a temporary teacher at a school and wrote: 'I have given up the ambition to get to a university.' In 1902 he began working at the Swiss Patent Office in Bern, where he remained until 1909. In 1896 he had renounced his German citizenship, though he did not become a Swiss citizen until 1901.

In 1905, still at the Patent Office, Einstein wrote three papers, any one of which was important enough to merit a Nobel Prize – though in fact this honour did not come his way until 1921. The first paper showed that electromagnetic energy was emitted in 'packets', or quanta, rather than continuously, and helped to lay the foundation of quantum mechanics. The second proposed what is today called the special theory of relativity, and that mass and energy were inextricably linked. The third dealt with statistical mechanics.

Any one of these papers would be enough to retire on, but Einstein was just beginning; in 1908 he became a lecturer at Bern University, and in the following year he resigned from the Patent Office to become professor of physics at the University of Zürich. In 1914, Einstein returned to Germany to take up important academic positions, though he retained his Swiss citizenship. Soon afterwards, in 1915, he published the general theory of relativity.

▼ **The Einstein Cross**
Light from a background quasar is bent around a foreground galaxy to form four images of the quasar.

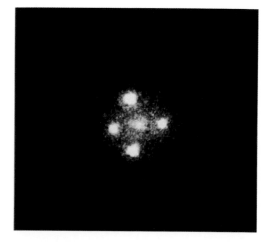

By now Einstein was world famous, and in 1919 one of his predictions based on relativity theory – the bending of light rays when passing a massive body – was verified by the British eclipse expedition in which Arthur Eddington was the main researcher.

Einstein spent most of the rest of his professional career arguing against the direction that quantum physics had taken. The theory is fundamentally based on probability – and Einstein's riposte was simple and direct: 'God does not play dice!'

Einstein first visited the United States in 1921, and other visits followed. In 1933, when the Nazis came to power in Germany, Einstein, as a Jew, wisely did not return there, and spent the rest of his life in the United States. He became a U.S. citizen in 1940. In 1952 he was invited to become the first president of Israel, but declined.

All his life Einstein worked for international peace. When one of us (PM) met him in 1939, he proved to be very much as had been expected. Only a week before he died, on April 18, 1955, he wrote a letter to the philosopher Bertrand Russell agreeing that his name should go on a manifesto urging all nations to give up nuclear weapons.

Among other things, Einstein was an excellent violinist. One story relates how he was playing a duet with a professional violinist, and made a mistake. The professional looked at him pityingly: 'The trouble with you, Albert, is that you cannot count!'

George Ellery Hale 1868–1938

George Hale was born in Chicago, and soon showed a great interest in astronomy. His parents were well off and were able to provide him with a well-equipped private observatory. He was concerned mainly with the Sun, and was the first to realize that sun spots are essentially magnetic phenomena, caused by the cooling of the bright solar surface when the underlying lines of magnetic force break through. But before long he turned his attention to the stars, which meant using very different techniques. With the Sun, there is plenty of light to spare; with the stars, it is essential to collect every scrap of light available.

▶ **Hale 200-inch telescope**

The telescope in Palomar, California, is photographed against the backdrop of the Milky Way.

Hale's frequent call was for 'More light!' and for this he needed larger and larger telescopes. Having drawn up plans for the instruments he needed, he had an amazing knack of persuading friendly millionaires to finance them. His methods were sometimes unconventional. On one occasion he and a potentially useful millionaire were at the same function, but at different tables. Hale strode into the dining hall and switched labels, putting himself next to the millionaire. By the time that the cheese and biscuits were served, the millionaire had agreed to finance a giant telescope!

Hale was the driving force behind four major instruments: the 40-inch refractor at the Yerkes Observatory in Wisconsin (still the largest refractor in the world, and likely to remain so), completed in 1897; the Mount Wilson 60-inch reflector, brought into use in 1908; the Mount Wilson 100-inch, completed in 1917, both near Pasedena, California; and the Mount Palomar 200-inch, also in California, which saw 'first light' in 1948.

It is sad that Hale did not live to see it. For many years the 200-inch, now called the Hale Telescope, was in a class of its own, and almost at once it demonstrated its vast light-gathering power. Walter Baade used it to examine distant Cepheid variables in external galaxies, proving that the galaxies were twice as far away as had been thought.

Times have changed, and most new observatories are collaborations between governments and the result of international treaties rather than a good dinner. But without George Ellery Hale, researches into the nature and evolution of the Universe would have proceeded much more slowly.

▲ Fred Hoyle

Sir Fred Hoyle with Patrick Moore, photographed in 1982.

Fred Hoyle 1915–2001

Fred Hoyle – he was christened Fred, not Frederick or Alfred – was one of the most influential and controversial astronomers of modern times. On occasion he could be wrong when most other people were right – but he could also be right when almost all others were wrong!

Hoyle was born in Bingley, in Yorkshire, son of a wool merchant. He attended the local grammar school, and gained a scholarship to Emmanuel College, Cambridge, where he read mathematics. He was drawn to astrophysics largely because of his association with Raymond Lyttleton, with whom he wrote papers about accretion and stellar evolution.

During the Second World War, Hoyle worked on radar, together with Hermann Bondi and Thomas Gold; when off duty they discussed astronomy. In 1948 after the war's end, Bondi and Gold formulated the steady-state cosmology, and Hoyle was an immediate supporter and collaborator. He never abandoned his belief even when it was shown that, in its original form at least, the idea of continuous creation was untenable.

From 1945 Hoyle was based at Cambridge, and in 1958 he became Plumian Professor of Astronomy. It was at Cambridge that he produced his most brilliant work, and for some years he was generally regarded as the world's leading astrophysicist; much of our current knowledge about stellar evolution and constitution is due to him. There is little doubt that he ought to have shared the 1983 Nobel Prize with his co-author Alfred Fowler, and he made no secret of his annoyance at being passed over.

His tireless advocacy and fund-raising led, in 1966, to the establishment of the Institute of Theoretical Astronomy. The institute quickly established an international reputation; Hoyle used to attract visiting academics for a summer or two with a promise of free photocopying! Its building is now named after him. Unfortunately his relationships with his colleagues were not always happy, and he left Cambridge abruptly in 1972, retiring to his home in the Lake District, where he stayed until finally moving to the sedate atmosphere of Bournemouth, on England's south coast. He produced many more books, ranging from novels to discussions of panspermia and palaeontology; it must be said not all of these enhanced his reputation, and at least one, dealing with his claim that the famous fossil archaeopteryx was a fake, is best forgotten.

As a broadcaster he was superb; his first BBC series, *The Nature of the Universe*, achieved tremendous popularity and influence, although it proved so controversial that a special commission, incorporating among others the Astronomer Royal and the Archbishop of Canterbury, had to be convened to decide if the broadcasts should go ahead. In 1957 he wrote a classic science-fiction novel, *The Black Cloud*, and followed it up with others, including *A for Andromeda*, which became a successful TV series.

Hoyle died in Bournemouth in 2001, mourned by his friends and colleagues all over the world. Even those who disagreed with some of his views would never question his brilliance, his originality and his integrity.

Edwin Powell Hubble 1889–1953

Edwin Hubble is widely regarded as the greatest American astronomer of all time. He was born at Marshfield in Missouri, and during his schooldays distinguished himself not only in his academic studies but also by his athletic prowess; he was a skilled amateur boxer and baseball player. He read law, and won a Rhodes Scholarship to Oxford University in England. He enjoyed his time there and always retained something of his 'Oxford accent'. When the United States joined in the First World War, Hubble enlisted in the army, and rose to the rank of major. Rather to his disappointment he did not see active service, but was thereafter glad to be addressed as Major Hubble.

Hubble decided that he did not wish to remain in law, claiming that he'd rather be a third-rate astronomer than a first-rate lawyer. After graduating he was able to use the 100-inch Hooke telescope at Mount Wilson. He looked for Cepheid variables in some of the spiral galaxies, and found them. He determined their periods, and hence their distances, showing beyond doubt that they were far too remote to be members of our Galaxy; they simply had to be external systems. From its Cepheids, Hubble gave the distance of the Andromeda Spiral as 900,000 light-years, later reduced to 750,000 light-years. This was a marked underestimate – there was an error in the Cepheid scale, which did not come to light until Walter Baade's work in the 1950s – but the vital point had been made.

At the Lowell Observatory in Arizona, Vesto Slipher had found that apart from a few nearby systems (those of what we now call the Local Group) all the galaxies are receding from us. Hubble found that distance was linked with recessional velocity: the greater the distance the greater the speed. The speeds could be measured spectroscopically by using the Doppler effect; the entire universe was expanding. During all this work Hubble was ably assisted by Milton Humason, who began his career as a mule driver up the road to the Mount Wilson Observatory, and ended it as one of the world's most eminent and respected astronomers.

Humason lived until 1972, Hubble died in 1953, just as the then-new Palomar reflector was showing that the earlier work on the distance of galaxies needed to be revised. He was active until almost the end of his life. Today we refer to the Hubble Constant, which relates the speed of recession of the galaxies to their distance from us,

currently measured as about 69 kilometres per second per megaparsec, and, of course, the first great space telescope was named in his honour.

Gerard Kuiper 1905–1973

Gerard Kuiper, Dutch by birth, spent most of his career in North America and took US citizenship in 1933. Kuiper was educated in his native Netherlands, but then emigrated. He was director of the Yerkes Observatory, Wisconsin, from 1947 to 1949 and again from 1957 to 1960, when he left to become the first director of the Lunar and Planetary Laboratory at Tucson, Arizona. He was still in office at the time of his death 13 years later.

Kuiper was concerned mainly with the Solar System; he discovered Neptune's second satellite, Nereid, and in 1944 showed that Titan, the largest satellite of Saturn, has a dense atmosphere. His name is attached to the Kuiper Belt of small bodies out beyond Neptune; Pluto was the first of these bodies to be discovered, but there are now over a thousand recorded.

Despite his successes he was well aware that conditions of seeing on the Earth's surface leave a great deal to be desired, and the best results are obtained from high altitudes – in fact, mountain tops. Working on the site of Mauna Kea in Hawaii, he noticed that the last thing he could see whenever clouds closed in was the summit of Mauna Kea sticking up above the cloud. This dormant volcano, on the Big Island, has a peak some 14,000 feet (4300 metres) above sea level, and if you go there you will find that your lungs are taking in only 39 per cent of the normal amount of oxygen. This can be dangerous: different people react in different ways, and there are some who cannot tolerate the conditions at all. Moreover, working at such an altitude is not easy. One's thinking processes are markedly slowed down.

▲ Bernard Lovell

Sir Bernard Lovell being interviewed by Patrick Moore in 1968.

There were no roads up Mauna Kea, and the whole area was totally barren. All the same, Kuiper was anxious to build an observatory there. At first he had few supporters, but in the end he had his way, and today the summit of the volcano bristles with telescopes – including the Keck reflectors; UKIRT, the United Kingdom Infrared Telescope; and JCMT, the James Clerk Maxwell Microwave Telescope. Generally speaking the seeing conditions are excellent; Kuiper had been absolutely right.

Kuiper was also one of the main supporters of the early space missions to the planets, and he was involved in planning the orbits of the spacecraft. He lived to see the Moon landings, though, sadly, he died before the main planetary programmes were under way.

Mauna Kea is a fascinating place; it is within sight of its twin Mauna Loa and neighbouring Kilauea, which are two of the most active volcanoes in the world. On one occasion lava from Mauna Loa reached the outskirts of the town of Hilo and was halted only at the last moment. Mauna Kea itself is dormant. At least, we hope so: an eruption there would do the great observatories no good at all.

Bernard Lovell 1913–

Bernard Lovell – he never used his other two Christian names, Alfred and Charles – was born in Gloucestershire, and began his scientific career as a physicist, graduating from Bristol University. He became a specialist in cosmic-ray research at Manchester University, but when war broke out in 1939 he became involved with work at the Air Ministry, and made very valuable contributions with the use of radar for detection and navigational purpose. At the end of the war this work was put to astronomical use, and Lovell obtained an ex-army mobile radar unit for his cosmic-ray studies, but interference from electric trams in the city caused him to move his equipment to Jodrell Bank, in the surrounding countryside. Initially there was a dispute between the two farmers who were selling the land, and on one occasion the scientists found themselves faced with a decidedly short-tempered bull!

Lovell turned his attention to detecting meteor trails by using radar, and this led on to greater things. In 1951 he was appointed professor of radio astronomy at Manchester University. He planned a large steerable radio telescope, and after many trials and tribulations succeeded in having it built. Inevitably it cost much more than originally proposed; Lovell took risks – at one stage it was even suggested that he might be sentenced to imprisonment! Typically, at the height of the crisis, when he was summoned to meet some very important officials from the government, he was busy playing cricket – with great success; he was an excellent cricketer.

The financial situation was saved by the launch of the first Soviet artificial satellite, Sputnik I. Outside Russia, only the newly-completed 250-foot Jodrell Bank 'dish' was capable of tracking it, and almost overnight Lovell was transformed from a reckless spendthrift to a national hero. Although this has only ever been a minor part of its work, the giant dish is still used for satellite tracking occasionally. It detected signals from the European Space Agency Huygens probe as it landed on Titan in early 2005.

The Jodrell Bank dish marked the start of the modern phase of radio astronomy, and it was only right to rename it the Lovell Telescope on the occasion of its 30th anniversary. The inscription '1956–1986' was painted on the dish – without Sir Bernard's knowledge; he had to be kept out of view of the telescope until the unveiling ceremony. (One of us – PM – was actively involved in this plot!)

Though the 250-foot is no longer the world's largest radio telescope, its influence has been paramount. Without Lovell's resolution and energy it would not have been built and it is fair to regard him as the 'Isaac Newton of radio astronomy'.

Martin Rees 1942–

Martin Rees grew up in Shropshire, but graduated from Cambridge and has spent almost the whole of his career there. He has been at the forefront of research into the nature of black holes, and has been responsible for major advances in our understanding of quasars, gamma-ray bursts, formation of galaxies and the Cosmic Microwave Background. In fact he has contributed to almost every branch of astrophysics and cosmology.

He has always had a special interest in the nature of compact objects; for example he predicted (correctly) that massive black holes would be found near the centre of galaxies, including ours. It was also he who was one of the first to establish the nature of the strange, super-violent explosions known as gamma-ray bursts.

In his attempts to explain how the Universe emerged from the Dark Ages, he has examined how the first generation of stars, quasars and galaxies formed then ionized much of the Universe. He made the first predictions about polarization and other detailed characteristics of the Cosmic Microwave Background.

Quite apart from all this, Rees is a lecturer and broadcaster, and in writing popular books he has few, if any, equals: he has the ability to write about very difficult subjects and make them sound straightforward and this has been of enormous value in his stint as Astronomer Royal from 1995 to 2005. He has always been very enthusiastic about international collaboration in science, and when the British government planned to destroy the Royal Greenwich Observatory, which had been founded by order of King Charles II, Rees did everything in his power to save it. Unfortunately, his efforts were not successful; the civil service had made up its mind.

▲ Lord Rees of Ludlow
The then Sir Martin Rees and Patrick Moore at the opening of the South Downs Planetarium in Chichester, England.

◄ Quasars
PG 0052+251 is 1.4 billion light years from Earth, at the core of a spiral galaxy. Lord Rees is at the forefront of research into the nature of Quasars, and has played a major role in formulating our current understanding of these remote objects.

▶ **Field of wires**

Martin Ryle was the driving force behind the development of the radio astronomy group in the Cavendish Laboratory in Cambridge, a tradition which continues today. The most notable contributions of the Cambridge group include the development of the first systematic catalogues of radio sources and the discovery of pulsars. This dramatic – and unexpected – result came from this telescope, the 4-acre array. Ryle and his colleague, Anthony Hewish, were awarded the 1974 Nobel Prize in Physics for their work.

It is evident that Rees has had a tremendous impact in astronomy, not only by his own personal research but in his encouragement of others. Now elevated to the House of Lords, he has become president of the Royal Society.

Martin Ryle 1908–1984

Martin Ryle, one of the leading radio astronomers of the 20th century, is best known for his development of revolutionary radio telescopes and using them to make measurements of the most remote galaxies then known. He was born in Brighton, graduated with his degree in physics from Oxford University in 1939 and then, during the Second World War, was involved in the development of radar.

He then went to the Cavendish Laboratory at Cambridge, and it was at Cambridge that he spent the rest of his scientific career. While at Cambridge, Ryle supervised the preparation of catalogues of radio sources. The 3C survey of 471 sources in 1959 is still used as a major reference source of data, and the 4C survey listed 5000 radio sources. Among the most important discoveries of this era was the definite detection of differences between the local and distant regions of the Universe, which formed the first real observational support for the Big Bang theory, of which Ryle was an enthusiastic advocate.

In 1972 Ryle succeeded Sir Richard Woolley as Astronomer Royal: he was the first Astronomer Royal based away from the Royal Greenwich Observatory (RGO), and it has been said that this departure from tradition led indirectly to the demise of the RGO before the end of the century – a loss that can only be described as bureaucratic vandalism. However, this was most certainly not Ryle's fault. After his retirement as Astronomer Royal, in 1982, he devoted his time to social and environmental issues.

Karl Schwarzschild 1873–1916

Schwarzschild was born in Frankfurt, Germany. His father was a wealthy businessman and his boyhood was pleasant and peaceful. He became interested in astronomy at an early age, and saved up enough money to buy a small telescope. His first two papers, on the theory of orbits of double stars, were published when he was only 17.

He studied at the University of Strasbourg, and then obtained his doctorate from Munich University. He published several important papers about stellar astronomy, and also made himself known as a splendid lecturer. It was said that he had the knack of making difficult problems sound easy. He was a professor at Göttingen University, and became director of the observatory there. He then went on to become director of the Astrophysical Observatory at Potsdam, the most prestigious post available to any astronomer in Germany. He was a great success, and it was at this time that he made important contributions to spectroscopy.

Then came the First World War. Patriotically Schwarzschild volunteered for active service, and after taking charge of a weather station in Belgium, he went to the Russian Front to calculate missile trajectories. Here, while a serving soldier in constant danger, he wrote two basic papers, one on quantum theory and the other on Einstein's theory of relativity. The relativity paper gave the first exact solution of Einstein's general gravitational equations, leading on to the understanding of the geometry of space close to a massive body. He sent the paper to Einstein, and the reply has been preserved: 'I had not expected that anyone could formulate the exact solution of the problem in such a simple way.'Those two papers formed the basis for later studies of black holes.

Sadly, Schwarzschild was given no chance to do more. The harsh conditions on the Russian front made him contract a painful skin illness for which at that time there was no known cure. He was invalided out of the army in 1916, to no avail; he died two months later – a victim of one of mankind's senseless wars.

He was not forgotten: in 1960 the Berlin Academy of Sciences officially named him as the greatest German astronomer of modern times. A major observatory at Tautenberg is named after him. His son, Martin, also became a distinguished astronomer.

Harlow Shapley 1885–1972

Harlow Shapley made many important contributions to astronomy, but will always be best remembered as the man who measured the size of the Galaxy and showed that, far from being centrally placed, the Sun is well out towards the edge of the main system.

Shapley was born at Nashville in Missouri. His father was a farmer, and his early education was somewhat sketchy. At the age of sixteen, Shapley left school and became a newspaper reporter in Kansas. He intended to make journalism his career, but when he tried to enrol at the University of Missouri's School of Journalism he found that he would have to wait for a full year before a vacancy occurred – so he joined an astronomy course instead. In fact, he became an astronomer purely by accident.

After graduating Shapley went to Princeton, where the head of the astronomy department was Henry Norris Russell. Shapley received his doctorate and then joined the staff of Mount Wilson Observatory in California, where he remained for seven years, probably the most productive of his entire career. It was here that he concentrated upon the size and shape of the Galaxy. He knew the position of the globular clusters in the sky – most are in the southern hemisphere – and by using their Cepheid variables as 'standard candles' he could work out their distances. Since the globulars lie around the main Galaxy, this gave him the size of the system itself. His value was over-estimated, for he could not take into account the 'interstellar fog', the absorption of light by thinly-spread matter in space; but he had taken the essential step.

In another respect he was wrong. Originally he believed that the 'spiral nebulae' were parts of our Galaxy rather than external systems or 'island universes', and this led to a famous debate with the astronomer Heber D. Curtis. It may be said to have ended in an honourable draw. Shapley was right about the size of the Galaxy; Curtis was right about the nature of the spirals.

Shapley left Mount Wilson in 1920 to become director of the Harvard College Observatory, where he remained until retiring in 1952. He wrote books, both technical and popular, and became active in administration.

One of us (PM) has an endearing recollection of Shapley. In 1966 he took part in an episode of *The Sky at Night* together with another eminent astronomer, Baret Bok. In those days it was essential to use film inserts to the main programme, and we were sitting in a row: Shapley, Moore, Bok. During a recording Shapley and Bok changed places, with malice aforethought. Nobody noticed – until the programme was transmitted, when they moved round from chair to chair in a most bewildering manner!

Jakov Borisovich Zeldovich 1914–1987

Jakov Zeldovich was the greatest member of the Soviet school of cosmologists that developed during the 1950s and 1960s. He was born in Minsk, formerly in the Soviet Union, and now the capital of Belarus.

From the age of 17, Zeldovich worked at the Institute of Chemical Physics of the Academy of Sciences of the USSR, in Leningrad (now restored with its original name of St Petersburg), where he essentially educated himself. Somehow he managed to escape Stalin's purges, in which so many eminent researchers were killed, and became a professor at Moscow State University. He produced important papers about shock waves and gas dynamics, and played a significant role in the development of Soviet nuclear and thermonuclear weapons.

In the 1950s Zeldovich turned to nuclear physics and the theory of elementary particles. It was not until the 1960s that his attention was centred mainly on astrophysics and cosmology; he became head of the division of relativistic astrophysics at the Sternberg Astronomical Institute in Moscow. He wrote several papers dealing with the dynamics of neutron emission during the formation of black holes, the formation of clusters and galaxies, and the large-scale structure of the Universe.

With Rashid Sunyaev, Zeldovich proposed what is now called the Sunyaev-Zeldovich Effect, an important method for detecting galaxy clusters, which has the advantage of being equally effective in the distant and nearby Universe. Several telescopes are now being built to take advantage of his – then theoretical – insight.

Perhaps Zeldovich's most influential suggestion was the realization that a very massive cloud of matter should not collapse uniformly, but would develop instabilities leading to an asymmetric shape. The resulting 'pancake distribution' is indeed observed in clustering of galaxies and is significant in considerations of the large-scale structure of the Universe.

Zeldovich was a pioneer in attempts to relate particle physics to cosmology, and to develop a theory that unified quantum mechanics with the theory of gravity. Quite apart from all this, he was an outstanding teacher, and the leader of one of the world's leading research groups in general relativity and cosmology. He received many honours, including the I.V. Kurchatov Gold Medal of the Soviet Academy of Sciences (1977), the Soviet Government Order of Lenin (1962 and 1971) and the Bruce Medal (1983). Zeldovich lived through many changes in what is now Russia, but apparently his scientific career was more or less uninterrupted. He died in Moscow in 1987.

Fritz Zwicky 1898–1974

Zwicky, one of the most brilliant and most extraordinary astronomers of modern times, was born in Bulgaria, but his parents were Swiss, and he always kept his Swiss nationality. He graduated from the University of Zürich, and then, in 1925, emigrated to the United States to join the California Institute of Technology. He remained there permanently, becoming a professor of astrophysics in 1942 and holding this post until his retirement in 1968.

▲ **Fritz Zwicky**

Zwicky's first major contribution to astronomy concerns the life stories of very massive stars. He realized that when their nuclear 'fuel' ran out they would explode violently, and it was he who first coined the term 'supernova'. He knew that supernovae in our Galaxy were rare, but he calculated that there ought to be one supernova every 200 to 400 years, which means we are now overdue for our first galactic supernova since the discovery of the telescope, which is frustrating to say the least!, and the same would be true of other galaxies. He was able to use one of the main telescopes on Mount Wilson to search for them, and by 1936 he had discovered 36 supernovae – far more than most people had expected. With Walter Baade, he also suggested that after a supernova outburst, the remnant of the star would collapse into a small, incredibly dense globe made only of neutrons. This idea was also received with some scepticism, but of course proved to be correct.

Zwicky also studied the ways in which stars and galaxies move, and he realized that a galaxy cluster will soon disperse if not 'glued' together by matter that we cannot see; in fact this was the first mention of the 'dark matter' that modern astronomers have come to believe dominates the Universe.

All this was brilliant research but Zwicky was an unusual sort of person, and to say that he was irascible is to put it mildly; anyone who disagreed with him was automatically regarded as a mortal enemy. He was immensely strong, and made a habit of doing handstands in the dining hall of the observatory to make sure that everybody was aware of it. He referred to his colleagues as 'spherical bastards' – spherical because he maintained that they appeared to be bastards no matter from which direction you looked at them. He was convinced that others were stealing his ideas without giving him credit, and he was particularly venomous about Edwin Hubble. On one occasion he opened the slit of the observatory dome and ordered his night assistant to fire bullets through it, claiming that this would improve the seeing (it did not). We have already noted his opinion of Walter Baade!

When Zwicky retired it is not likely that many of his colleagues were sorry to see him go. But at least he had made outstanding contributions to astronomy, and these were of the greatest value to the 'spherical bastards' who followed him.

▶▶ **Image (overleaf):** Thousands of galaxies can be seen in this, the most penetrating view of the Universe yet, the Hubble Ultra Deep Field.

TIME LINE OF THE UNIVERSE

Time after the Big Bang (A.B.)	Event	Years ago / future
0	Big Bang	13.7 billion years ago
10^{-35} to 10^{-33} seconds	Inflationary period	
10^{-33} seconds	Birth of quarks and antiquarks. They annihilate each other, leaving a slight excess of quarks.	
10^{-5} seconds	Quarks combine to form protons and neutrons.	
10^{-3} seconds	Formation of hydrogen and helium atoms	
1 to 3 minutes	Formation of light elements up to boron	
370,000 years	Emission of CMB – Universe becomes transparent	
200 million years	Birth of first stars, reionization	13.5 billion years ago
3 billion years	Formation of mature galaxies, quasars, and the oldest stars in Milky Way	10.4 billion years ago?
9.1 billion years	Our Solar System including Earth formed	5.6 billion years ago
9.9 billion years	First fossils formed	3.8 billion years ago
13.4 billion years	First reptiles	320 million years ago
13.5 billion years	Africa splits from America; the dinosaurs appear	200 million years ago
13.64 billion years	End of the dinosaurs; small mammals diversify	65 million years ago
13.695 billion years	Primates including the first apes evolve.	5 million years ago
13.6998 billion years	Homo sapiens.	195,000 years ago
13.6999 billion years	End of last ice age, dawn of modern world	10,000 years ago
13.7 billion years	**Present day**	
14.7 billion years	Earth becomes uninhabitable.	1 billion years hence
18.7 billion years	Sun a red giant, destruction of Earth	5 billion years hence
23.7 billion years	Sun becomes a white dwarf	10 billion years hence
10^{14} years	Galaxy- and star-formation cease	A hundred thousand billion years hence
10^{36} years	50 per cent of all protons have decayed.	
10^{40} years	All protons gone, black holes dominate	
10^{100} years	Black holes disintegrate	
10^{150} years	Photon age: Universe reaches extreme low-energy state?	

GLOSSARY

Atom The ancient Greeks believed that matter could be broken down into indivisible units, which they called atoms. The modern idea of an atom is remarkably similar; it has a nucleus made up of positively charged particles called protons and neutral neutrons, which is surrounded by low-mass, negatively charged electrons. Electrons and protons have the same charge, although with opposite signs, and so a neutral atom must have an equal number of both. Carbon has 12, for example, while an atom of the lightest element – hydrogen – consists of a solitary proton and a single electron. Classically, physicists thought of the electron orbiting the nucleus just as the planets orbit the Sun, but in quantum physics things are much less straightforward.

Antimatter Modern theories of particle physics predict that every type of particle has an antiparticle, which has the opposite electrical charge but is otherwise identical. Collectively these antiparticles are known as antimatter. For example, the electron's antiparticle is the positron. When particle and antiparticle collide, they annihilate, releasing energy and (at least in science fiction) powering spaceships. In the very early stages of the Universe's evolution there were equal amounts of matter and antimatter, and it is unclear how we ended up in a Universe made overwhelmingly of matter.

Baryon A baryon is a particle that is composed of quarks. Examples include quarks themselves, neutrons and protons. Astronomers use the term 'baryonic matter' to distinguish the ordinary material in the Universe from the mysterious dark matter.

Billion Throughout this book we have used the standard scientific definition of billion, which is a thousand million (1,000,000,000 or 10^9). The older, English definition according to which a billion equalled a million million is now almost obsolete.

Black body An idealized emitter and absorber of radiation, to which a star approximates quite well. Any hot body emits electromagnetic radiation, and the spectrum of this emission for a black body is entirely determined by its temperature. A plot of energy against frequency (or colour) for a black body produces a smooth 'hump-backed' shape (see p.58) with a maximum intensity that moves to a higher frequency as the temperature increases. If a metal object such as a poker is heated, it appears first red, then orange, then yellow and eventually white hot. Similarly, the hottest stars have a blueish-white colour.

Black hole A black hole is a body that has a strong enough gravitational field to prevent even light – the fastest thing in the Universe – from escaping. The radius within which a particular mass must be confined to form a black hole is called the Schwarzschild radius. While considered for many years to be theoretical curiosities, there is now strong evidence that black holes exist. It seems that most galaxies have a supermassive black hole (as massive as several million suns) at the centre.

Charge Electric charge is a property shared by particles such as quarks, protons and electrons. Positive and negative electric charges attract each other, and it is this force that keeps the negative electrons bound to the positive nucleus in a neutral atom.

Comet An icy body, best described as a 'dirty snowball'. Comets are believed to form in the outer regions of the Solar System, where they remain in a 'reservoir' of comets known as the Oort Cloud until disturbed. Gravitational disturbances, such as the passing of a nearby star, can alter the orbit of a comet so that it swings into the inner Solar System. As it nears the Sun, the ice melts, and the familiar tail (which always points away from the Sun) forms. Although some comets are thrown out of the Solar System altogether, most become periodic and make regular visits to our neighbourhood. The first of these to be identified was Comet Halley, which last passed through the inner Solar System in 1985/86 and has a period of 76 years. Recent spectacular comets have included Hyakutake and Hale-Bopp in the late 1990s, and Comet Shoemaker-Levy 9, which crashed into Jupiter in 1994.

Constellation A group of stars that appear near each other in the sky, forming a recognizable pattern. The stars in a constellation have no physical link to each other and may be many thousands of light years apart. Although ancient cartographers chose to create their own constellations, in 1930 the International Astronomical Union selected 88 for their official list. Many of the best known groups – such as the Plough, which is merely part of Ursa Major – are not official constellations but are known as asterisms. The largest constellation is Hydra, the Sea Serpent, and the smallest is Crux, the Southern Cross. Although the constellations appear constant on the timescales of our human lifetimes, the stars are slowly moving and over time their familiar patterns will disappear.

Dark matter Over the last fifty years astronomers have come to realize that most of the matter in the Universe is composed not of ordinary atoms and molecules, but is in the form of some exotic dark matter. In fact, over 80 per cent of the mass in the Universe is made up of dark matter, which interacts with 'normal' (or baryonic) matter almost exclusively through gravity. Particle physics predicts the existence of a family of weakly interacting massive particles (or WIMPs), which may make up the dark matter, but this theoretical prediction has yet to be confirmed by either observation or experiment. Although the identity of dark matter has yet to be confirmed, observations allow scientists to constrain its properties; the majority of the evidence suggests that the correct model is one involving 'cold dark matter' – slow moving, massive particles.

Dimension A coordinate necessary to specify one's position. In everyday life we are used to three dimensions – length, width and height – along with time, which should also be considered a dimension. Some of the more exotic theories of particle physics suggest that there may be other 'hidden' dimensions, which only reveal their presence through the results of high-energy experiments.

Doppler effect The Doppler effect is most familiar from the change of pitch heard in a siren as an ambulance drives past. The waves emitted from an approaching source are compressed, and hence appear to have a higher pitch than those from a stationary source. Conversely, waves emitted from a receding source are stretched and hence appear to have a much lower pitch. The greater the relative speed between the source and the observer, the greater the shift in wavelength. The same effect applies to light; the stretching of the light means that light from a receding source will appear reddened, and that from an approaching source will appear shifted toward the blue end of the spectrum. Observations of galaxies beyond our own Milky Way show that all but the closest have spectra that are shifted toward the red end of the spectrum, revealing that they all appear to be receding from us. The further away the galaxy is, the faster it appears to be moving away from us, and it was this discovery that provided the first observational evidence for an expanding Universe.

Ecosphere The region in a Solar System within which the temperature is such that liquid water could exist on the surface of a rocky planet and therefore – assuming, of course, that all life needs water as life on Earth does – the region within which life could exist. In our own Solar System, Venus is closer to the Sun than the ecosphere and currently (although things could have been different in the past) Mars lies further away. Searches for planets around other stars are not yet sensitive enough to identify Earth-sized planets, but at least one Jupiter-mass object has been found within its system's ecosphere. Liquid water may therefore exist on any large moons orbiting that distant planet and potentially support life.

Electromagnetic radiation Visible light is just one part of the spectrum that stretches from ultra-high energy gamma rays and X-rays through the ultraviolet, then through the visible into the infrared, microwaves and then finally the radio wavelengths. Electromagnetic radiation in all of these forms is composed of electric and magnetic components, which move at the speed of light (see p.44).

Electron A low-mass particle (weighing the same as less than one thousandth [0.001] of a proton) with unit negative electric charge. Unlike protons and neutrons, electrons are not made up of quarks and appear to be truly 'fundamental' particles that cannot be broken into smaller parts.

Energy The law of conservation of energy (otherwise known as the first law of thermodynamics) is one of the most fundamental of all physical laws. It states that energy is neither created or destroyed, but can only be converted from one form to another. The famous equation $E=mc^2$ simply states that mass can be converted into energy or, equivalently, that mass is simply another form of energy. The nuclear reactions at the centres of stars convert mass into radiation and thermal energy.

Equator The imaginary circle drawn on a sphere so as to be an equal distance from both poles. We are relatively familiar with the Earth's equator, and the projection of this line on to the sky defines the celestial equator. It is useful as a point of reference for our coordinate systems, but its position on the sky has no physical significance.

Galaxy From the Greek word for 'milk,' the term galaxy was first applied to the Milky Way, which was seen as a bright strip of starlight running through the sky. Once it became clear that our own Galaxy was only one of many billions, the term was applied to mean any large group of stars and other material that exists as an independent system, held together by its own gravity. The two major classes of galaxies are the ellipticals and the spirals. The ellipticals are large, spherical systems of old stars with relatively little gas remaining to be converted into stars. By contrast, a spiral is characterized by a disk containing spiral arms that mark ongoing star-forming activity surrounding a central,

older bulge. It was believed for many years that elliptical systems formed from the collision of two spirals, but the process appears to be more complicated than that.

Gamma-ray burst The most powerful explosions in the Universe, these rapid events were first detected by satellites that were monitoring the Earth for signs of hidden testing of nuclear weapons during the Cold War. At least some of them are associated with extreme supernova outbursts known as hypernovae but others may be due to exotic phenomena such as the collision of black holes and neutron stars. As they are extremely luminous, they are visible even in parts of the distant Universe.

Gravity Although gravity is intrinsically the weakest of the fundamental forces, it is the only one of the traditional forces that acts on astronomical scales. Of the others, the strong and weak nuclear forces are extremely weak over large distances, and the electromagnetic force from positive and negative charges cancels out. The gravitational attraction between two objects is proportional to their masses and inversely proportional to the square of their separation. In other words, two masses moved so that the distance between them is halved will attract each other four times as strongly. The first systematic theory of gravity was due to Sir Isaac Newton, whose theories were expanded by Albert Einstein in his general theory of relativity.

Heat The scientific definition of temperature is rather different from the everyday one. The higher the temperature of a gas, the faster the atoms that compose it are moving. By contrast, 'heat' is usually used to mean the quantity of thermal energy present. For example, a firework sparkler is at a much higher temperature than a red-hot poker, but because there is much less mass in a firework sparkler than a poker there is more heat in the poker – which is why one can hold a sparkler but would be reluctant to hold a glowing poker.

Inflation An extension of the standard Big Bang theory that suggests that the Universe expanded at a greatly increased rate for a short period less than a second after the Big Bang. While direct evidence for inflation and a theoretical understanding of its causes has so far proved elusive, it provides an elegant solution to several observational problems with the standard Big Bang theory.

Ionization Energetic photons can knock electrons away from their atomic nuclei, which are then said to be ionized. In the energetic conditions immediately after the Big Bang, the electrons had too much energy to be captured by atomic nuclei and the entire Universe was ionized. As the Universe expanded, it cooled until the electrons could be captured by the nuclei in order to form neutral atoms. Upon the advent of the first sources of light, the electrons were liberated once more at an epoch known, rather confusingly, as reionization.

Light-year The distance travelled by light in a year when passing through a vacuum, equivalent to 9.5×10^{15} metres or nearly six thousand billion miles. The Sun is eight light-minutes away – the light we see left the star eight minutes before – the nearest other star is 4.2 light-years away. The Sun lies 26,000 light-years from the centre of the Milky Way Galaxy, which is itself 100,000 light-years across. Bodies located 13 billion light-years away are seen as they appeared just after the Big Bang.

Luminosity The luminosity of a light source reflects the rate of emission of light. In other words, the luminosity of a star reflects its intrinsic rather than apparent brightness. The Sun appears much brighter than the other stars in the sky because it is close to us, even though many stars are much more luminous than our own rather ordinary star.

Magnitude The traditional measure of brightness for astronomical objects. The scale is rather confusing; the lower the number, the brighter the source appears. By definition, the bright star Vega has a magnitude of 0.0 and a difference of five magnitudes corresponds to a difference of 100 times in brightness. Vega is therefore 100 times brighter than a star with a magnitude of 5. In dark skies, the naked eye can see to a magnitude of 6 or so. These are apparent magnitudes, but it is also common to refer to absolute magnitudes. These reflect the luminosity of the source and are defined as the apparent magnitude the source would have at the standard distance of ten parsecs.

Mass There are two scientific definitions of mass. The first is the property of a body to resist acceleration; it takes more effort to push a car than a football. The second is the property of a body that defines the strength of its gravitational attraction; objects with more mass have a stronger gravitational pull. The two turn out to be equivalent so that the same definition of mass can be used for both. A common mistake is to confuse mass with weight. Weight is the force exerted on an object by gravity. When Neil Armstrong stepped out onto the Moon's surface, his mass did not change but his weight certainly did.

Meteor A shooting star, or meteor, is caused by the entrance of a small particle, usually no bigger than a grain of sand, into the Earth's atmosphere. The friction of the atmosphere causes the particle to burn up, leaving a short-lived rapidly moving trail across the sky, which we see as a meteor. Many of these dust grains are associated with comets; as the comet makes repeated passes through the inner Solar System, dust thrown off from the nucleus spreads out along its orbit. When the Earth's orbit intersects that of the comet, we see a meteor shower. Meteors from the same shower will appear to trace back to a single area of sky, known as the radiant – the effect is similar to standing on a motorway bridge and seeing the two parallel carriageways appear to merge in the distance. Of particular note are the two most famous showers; the Perseids, which have their maximum in August, are the most reliable annual shower while the Leonids produce spectacular meteor storms on a roughly 33-year period. During such a meteor storm, for a short period of only an hour or so, the rate may approach a meteor per second.

Meteorite A large, usually asteroidal body, which has survived entry through the Earth's atmosphere and landed on the planet's surface. Remarkably, there is no record of anyone being injured by a meteorite fall, although several cars have been badly damaged in recent years! Although the source of most meteorites is the Asteroid Belt (or other asteroid-sized bodies), a few are likely to have come from the Moon or Mars. The most famous of these Martian meteorites, ALH84001, includes structures that look intriguingly like terrestrial bacterial fossils, but the scale is different. There is the possibility that they are there due to contamination following the meteorite's arrival on Earth, but they remain the most tangible evidence that life might exist, or has existed, on Mars. The majority of meteorites are found today in Antarctica, where they stand out against the icy background.

Milky Way A luminous band of faint stars that crosses the sky, containing many nebulae and dust clouds in addition to stars. It is the projection of the disk of our own Galaxy, which is also known as the Milky Way, onto the celestial sphere.

Nebula From the Latin word for 'mist', 'fog' or 'cloud', the term 'nebula' is used in astronomy to refer to any visible mass of gas and dust. The most famous nearby example, the Orion Nebula, is a region in which stars are forming from condensing gas and dust. The newly formed stars are then able to light up the surrounding gas in what is known as a 'reflection nebula'. In the latter stages of a Sun-like star's life it will expel its outer layers, forming a 'planetary nebula'. The name comes from the often disk-like appearance in small telescopes, but is somewhat unfortunate as there is no association with planets or with reflection nebulae. Dark nebulae, composed of dust blocking out light from more distant sources, are also observed; the most famous of these is the Coal Sack in the southern constellation of Crux.

Neutrino Small, lightweight particles that are produced as by-products of the nuclear fusion that powers stars such as the Sun. For many years, it was believed that they might be massless, but it is now clear they have a certain amount of mass (although not enough to account for the required amount of dark matter). This in turn solved the long standing 'neutrino problem' in which the number of neutrinos observed from the Sun was much lower than expected by theory. The tiny mass allows the neutrinos to change between three 'flavours' – electron, mu and tau neutrinos – en route between the Sun and the Earth, and previous generations of detectors were sensitive only to the least massive flavour. The number of neutrino flavours can be predicted by the Big Bang theory, and provides an excellent test of the idea that the Universe began in a hot, dense state.

Neutron Neutrons are one of the two types of particles – both composed of three quarks – that make up atomic nuclei. They weigh almost the same as protons, but carry no electric charge. Under the extreme conditions of a supernova explosion, protons and electrons can combine to form neutrons, resulting in a dense neutron star being produced from the dying star's core. The maximum mass for a neutron star is believed to be around eight solar masses; any larger than this and collapse to a black hole is inevitable.

Nucleus The nucleus of an atom is made up of positively charged protons and neutral neutrons, and contains almost all of the mass of the atom. At the high temperatures and pressures in the centres of stars, electrons are too energetic to be captured by the positively charged nucleus; so it is atomic nuclei that combine in fusion to form heavier elements. The number of protons in the nucleus of an atom defines its type, so that hydrogen has a single proton, helium two, lithium three and so on.

Parsec A unit of distance equal to 3.26 light-years. Seen from a distance of 1 parsec, the Earth would appear in the sky just 1 arcsecond (or a 3600th of a degree) away from the Sun.

Planck time In quantum mechanics, this is the smallest possible 'unit' of time, equal to 5×10^{-44} seconds. Even if a clock was accurate enough to measure a smaller period, quantum mechanics makes that impossible. Whether this is a real feature of the Universe or a statement of the inadequacy of quantum mechanics remains to be seen.

Positron The antiparticle of the electron, a positron has the same mass as an electron but the opposite charge. Like all matter–antimatter pairs, an electron and positron will annihilate on collision producing only energy.

Proton A positively charged particle composed of three quarks; protons are one of the two components that make up atomic nuclei.

Pulsar A rapidly spinning neutron star (produced in a collapsing supernova) will produce radiation in a thin beam from near both poles. As the star rotates, so this beam sweeps, lighthouse-like, across the sky. If it happens to cross the Earth, we see a rapidly pulsing source. So regular are these pulses that the first detection was labelled 'LGM-1?', standing for Little Green Men-1! There is one known example of a double pulsar, and scientists are able to exploit the information from its pulses to provide stringent tests of the theory of general relativity.

Quantum The fundamental insight of quantum mechanics is that a particle cannot have any arbitrary amount of energy, but must have a whole number of small 'bricks' of energy. These building blocks are known as quanta ('quantum' is the singular form). In our everyday lives, the effects of this phenomenon are small, because a single quantum is an extremely small amount; on the scales familiar to atoms and molecules, however, things are very different.

Quark The particles that combine to form protons, neutrons and other, more exotic, particles, quarks are now believed to be fundamental particles. In other words, they cannot be split. They come in six 'flavours', whimsically named 'up', 'down', 'strange', 'charm', 'top' and 'bottom'. (There has been a recently unsuccessful movement to rename the last two 'truth' and 'beauty'.) Quarks are attracted to each other by the strong nuclear force, and have the remarkable property that they can never be isolated as single quarks. The strong force increases with distance, so if two quarks are pulled apart the force attracting them to each other actually increases!

Quasar The original definition of quasar, or quasi-stellar object, was a star-like source that appeared to be at a great distance. Decades of observations have revealed that they are in fact galaxies harbouring extremely massive black holes at their centres, which are in the process of consuming huge amounts of dust and gas. This in-falling material radiates as it falls toward the central black hole, and this powerful source of radiation is responsible for our ability to see quasars from the most distant parts of the Universe. More common in the distant past, it has recently been suggested that all galaxies may have experienced a 'quasar-like' phase, relaxing to become 'normal' galaxies only when the reservoir of material to feed the central black hole has been exhausted.

Redshift The movement of spectral features toward the red end of the spectrum of a receding source, due to the Doppler effect. Astronomers also use redshift as a co-ordinate of time; the present day is an equivalent to a redshift of 0, and the observed redshift increases as we look back toward the earliest stages of the Universe's evolution. The most distant source yet observed is at a redshift of 6.4, which is equivalent to just 870 million years after the Big Bang (and 12.9 billion light-years away).

Spectrum Electromagnetic radiation passed through a prism (or a fine grating) will split into its component wavelengths, an effect most familiar from the sight of a rainbow in the sky. This is known as a spectrum, and the relative intensities of different wavelengths can encode a huge amount of information about the object that emitted the light. In particular, a series of dark or bright lines known as spectral lines (see p.58) acts as a fingerprint for each of the elements present in the source, allowing astronomers to identify the composition of even the most distant objects. Sir Isaac Newton coined the word spectrum from the Latin for 'to see.'

Standard form Standard form is the name given to the scientific notation used for very small and very large numbers. It takes the form of a number between 1 and 10, and then a factor by which that number should be multiplied. Rather than writing one and a half million as 1,500,000 it would be represented as 1.5×10^6. The factor above 10 represents the number of zeros to be put to the right of the first number. Similarly, then, a billion is written as 1.0×10^9 and a millionth as 1.0×10^{-6}.

Steady-state theory A now discredited rival to the Big Bang theory, which held that the Universe was in a constant state of continued expansion and small-scale creation of matter.

Strong nuclear force The force that binds quarks together to form larger particles such as protons and neutrons. It increases with distance, so that as two quarks are pulled apart the force between them increases.

Supermassive A term usually applied to the black holes at the centre of galaxies. An exact definition is elusive, but it is usually taken to mean several million times the mass of the Sun.

Supernova When a large star has used up the fuel in its centre, it suddenly collapses. The increased pressure at the centre will lead to the formation of a dense remnant such as a black hole or neutron star, at which point the majority of the remaining material will rebound outward in a powerful explosion known as a supernova. Such an event can easily outshine all the other stars in a host galaxy for a period of a few weeks before gradually fading. Exceptions to this general picture are type 1a supernovae, which are produced in binary systems in which material from a large star is able to build up on the surface of a white dwarf before reigniting when a critical density is reached. Somewhat frustratingly, there has not been an observed supernova in the Milky Way since the invention of the telescope; the nearest was in the satellite galaxy the Large Magellanic Cloud in 1987.

Wavelength The distance between two crests of a wave. The wavelength of red visible light is 4.0×10^{-7}m, while radio waves can have wavelengths of many kilometres.

Weak nuclear force The force responsible for particular kinds of radioactive nuclear decay.

Wormhole A purely theoretical structure (so far) that would allow distant regions of space to be connected by a 'short cut'. It has been speculated that black holes might mark one end of such a pathway, with material that descends into them re-emerging in a 'white hole'.

INDEX (Numbers in **bold** indicate illustrations)

FURTHER READING

The Canopus Encyclopedia of Astronomy, Paul Murdin and Margaret Penston eds. (Canopus, 2004)

Fred Hoyle: A Life in Science, Simon Mitton (Aurum/ABC, 2005)

Atlas of the Universe, Patrick Moore (Philip's/Firefly, 2005)

The Dark Side of the Universe, Iain Nicolson (Canopus/Johns Hopkins University Press, 2007)

The First Three Minutes: A Modern View of the Origin of the Universe, Steven Weinberg (Basic Books, 1993)

The Elegant Universe: Superstrings, Hidden Dimensions, and the Quest for the Ultimate Theory, Brian Green (Random House, 2000)

The Photographic Atlas of the Stars: The Whole Sky in 50 Plates and Maps, H. J. P. Arnold (Institute of Physics, 1999)

The Infinite Cosmos: Questions from the Frontiers of Cosmology, Joe Silk (Oxford University Press, 2006)

The Cambridge Guide to the Solar System, Kenneth Lang (Cambridge University Press, 2003)

Life's Solution: Inevitable Humans in a Lonely Universe, Simon Conway Morris (Cambridge University Press, 2003)

Astrophotography: An Introduction to Film and Digital Imaging, H. J. P. Arnold (Philip's/Firefly, 2003)

Yearbook of Astronomy, Patrick Moore (Macmillan, 2005)

PICTURE CREDITS

Every effort has been made to acknowledge correctly and contact the source and/or copyright holder of each picture, and Canopus Publishing Limited apologises for any unintentional errors or omissions, which will be corrected in future editions of this book.

Preliminary pages

pp. 4–5: ESA, NASA, Davide de Martin and Edward W. Olszewski; p. 8: Gordon Garradd; p. 10: Patrick Moore; pp. 12–13 Richard Gray.

Introduction

p. 13: European Southern Observatory; pp. 14–15: Nik Szymanek; p. 16: NASA; p. 17: NASA; p. 18: H J P Arnold; p. 19: Courtesy NASA/JPL-Caltech; p. 20: European Southern Observatory; p. 21: NASA; p. 22: James Symonds

Chapter 1

pp. 24–5: Brian Smallwood; p. 27: Greg Parker, Noel Carboni, New Forest Observatory; p. 28: M.C. Escher's "Cubic Space Division" © 2006 The M.C. Escher Company-Holland. All rights reserved; p. 29: European Southern Observatory; pp. 30–1: James Symonds; p. 32: Nanoscale Science Laboratory, Cambridge; p. 34: Courtesy of Brookhaven National Laboratory; p. 35: G.T.Jones, Birmingham University / Fermi National Accelerator Laboratory; p. 36: W. Purcell (NWU) et al., OSSE, Compton Observatory, NASA; p. 37: James Symonds; p. 38: R. Williams (STScI), the HDF-S Team, and NASA; p. 39: James Symonds; p. 40: James Symonds; p. 41: Brian May.

Chapter 2

pp. 42–3: Brian Smallwood; p. 44: James Symonds; p. 45: NASA/WMAP Science Team; p. 46: NASA/GSFC/JPL-Caltech; p. 47: K Lanzetta (SUNY Stony Brook) and NASA; p. 48: NASA; p. 49: Royal Astronomical Society; p. 50: NSF; p. 51: Kamioka Observatory, ICRR (Institute for Cosmic Ray Research), The University of Tokyo; p. 52: BOOMERANG Collaboration; p. 53: Max-Planck-Institute for Astrophysics; p. 54: 2dFGRS Team; p. 55: James Symonds; p. 56: James Symonds; p. 57: SOHO; p. 58 (top): James Symonds; pp. 60–61: Courtesy NASA/JPL-Caltech; p. 62: NASA; p. 63: Dr. Christopher Burrows, ESA/STScI and NASA; p. 64: Anglo-Australian Observatory / David Malin Images; p. 65: European Southern Observatory; p. 66: Gamma-Ray Astronomy Team / NASA; p. 67: NASA and The Hubble Heritage Team (STScI/AURA); p. 68: NASA/ESA/R. Sankrit and W. Blair (Johns Hopkins University); p. 69: James Symonds.

Chapter 3

pp. 70–1: Brian Smallwood; p. 72: NRAO/AUI; p. 73: NASA, ESA, and S. Beckwith (STScI) and the HUDF Team; p. 74: E.J. Schreier (STScI), HST, and NASA; p.

75: COBE/DIRBE/Richard Sword; p. 76 (top): Axel Mellinger; p. 76 (bottom): Patrick Moore; p. 77 (top): ESO/S. Gillesan et al.; p. 77 (bottom): James Symonds; p. 78: NASA and The Hubble Heritage Team (STScI/AURA); p. 79 (top): Todd Boroson/NOAO/AURA/NSF; p. 79 (bottom left): GALEX, NASA; p. 79 (bottom right): Jason Ware / www.galaxyphoto.com; p. 80: The Hubble Heritage Team (AURA/STScI/NASA); p. 81: Mark Westmoquette (University College London), Jay Gallagher (University of Wisconsin-Madison), Linda Smith (University College London), WIYN//NSF, NASA/ESA; p. 82: James Symonds; p. 83: X-ray: NASA/CXC/CfA/M.Markevitch et al.; Optical: NASA/STScI; Magellan/U.Arizona/D.Clowe et al.; Lensing Map: NASA/STScI; ESO WFI; Magellan/U. Arizona/D.Clowe et al.; p. 84: NASA/ESA; p. 85 NASA/CXC/M.Weiss; p. 86: James Symonds; p. 87: © Bettmann/CORBIS; p. 88: NRAO/AUI; p. 89: NASA, Andrew Fruchter and the ERO Team [Sylvia Baggett (STScI), Richard Hook (ST-ECF), Zoltan Levay (STScI)] (STScI).

Chapter 4

pp. 90–1: Brian Smallwood; p. 92: European Southern Observatory; p. 93: NASA and The Hubble Heritage Team (AURA/STScI); p. 94: NASA and The Hubble Heritage Team (STScI/AURA); p. 95: NASA, ESA, STScI, J. Hester and P. Scowen (Arizona State University); p. 97: James Symonds; p. 98: IRAS; p. 99: European Southern Observatory; p. 100 (top): James Symonds; p. 100 (bottom): courtesy NASA/JPL-Caltech; p. 101 (top): NASA; p. 101 (bottom): courtesy NASA/JPL-Caltech; p. 102: NASA; p. 103 (top): ESA/DLR/FU Berlin (G. Neukum); p. 103 (bottom): NASA and The Hubble Heritage Team (STScI/AURA); p. 104 (left): courtesy NASA/JPL-Caltech; p. 104 (right): Ian Sharpe; p. 105 (left): Courtesy NASA/JPL-Caltech; p. 105 (right): Courtesy NASA/JPL-Caltech; p. 106 (top): Sebastian Deiries / ESO; p. 106 (bottom): Thomas Balstrup and Lars T. Mikkelsen; p. 107 (top): NASA/JPL/Space Science Institute; p. 107 (bottom): Kate Shemilt; p. 109: NASA/ESA.

Chapter 5

pp. 110–11: Brian Smallwood; p. 112: James Symonds; p. 113: Pete Lawrence; p.115 Patrick Moore; p. 116: OAR/National Undersea Research Program (NURP); NOAA; p. 117: Stephen Low Productions; p. 118: Brian May; p. 119: Virgil L. Sharpton, University of Alaska, Fairbanks; p. 120–1 (top): ESA; p. 121 (bottom): NASA/JPL-Caltech; p. 122 (top): Patrick Moore; p. 122 (bottom): Patrick Moore; p. 123: HiRISE, MRO, LPL (U. Arizona), NASA; p. 124: NRAO/AUI; p. 125: Anthony Holloway; p. 126: NASA/JPL-Caltech; p. 127: NASA

Chapter 6

pp. 128–129: Brian Smallwood; p. 130 (bottom left): Brian Smallwood; p. 130 (bottom right): Patrick Moore/Brian May; p. 131 (top): Landsat Pathfinder Project; p. 131

(bottom): www.asiafoto.com; p. 132 (top): Phil James (Univ. Toledo), Todd Clancy (Space Science Inst., Boulder, CO), Steve Lee (Univ. Colorado), and NASA; p. 132 (bottom): NASA; p. 134: Cassini radar mapper, JPL, ESA, NASA; p. 135 (top): A. Dupree (CfA), R. Gilliland (STScI), FOC, HST, NASA; p. 135 (bottom): James Symonds; p. 136: ESA & Garrelt Mellema (Leiden University, the Netherlands); p. 137: The Hubble Heritage Team (AURA/STScI/NASA); p. 138 (top): Bruce Balick (University of Washington), Vincent Icke (Leiden University, The Netherlands), Garrelt Mellema (Stockholm University), and NASA; p. 138 (bottom): NASA/ESA & Valentin Bujarrabal (Observatorio Astronomico Nacional, Spain); p. 139 (top): NASA; ESA; Hans Van Winckel (Catholic University of Leuven, Belgium); and Martin Cohen (University of California, Berkeley); p. 139 (bottom): Andrew Fruchter (STScI) et al., WFPC2, HST, NASA; p. 139 (top) NASA; p. 140 (bottom): NASA/SAO/CXC; p. 142: NASA/CXC/SSC/J. Keohane et al.; p. 143: J. M. Cordes & S. Chatterjee; p. 144: Image courtesy of NRAO/AUI and HST/STScI; p. 145: Brad Whitmore (STScI) and NASA; pp. 146–7: NASA, H. Ford (JHU), G. Illingworth (UCSC/LO), M.Clampin (STScI), G. Hartig (STScI), the ACS Science Team, and ESA.

Chapter 7

pp. 148–149: Brian Smallwood; p. 151: ESO; p. 152: Werner Benger; p. 153: Canada-France-Hawaii Telescope/J.-C. Cuillandre/Coelum; p. 155: Brian Smallwood.

Epilogue

p. 156: NASA

Practical Astronomy

p. 158: Pete Lawrence; p. 159: Kate Shemilt; p. 160: John Fletcher; p. 161: Jamie Cooper; p. 162: Ian Sharpe; p. 163 (top): Damian Peach; p. 163 (bottom): Brian May; p. 164: Jamie Cooper; p. 165: Damian Peach; p. 166 (top): Ian Sharpe; p. 166 (bottom): Patrick Moore; p. 167: Brian May; pp. 168–72: all star maps courtesy James Symonds; pp. 174–5: © Instituto de Astrofisica de Canarias.

Biographies

p. 176: ANTU/UT1 + FORS1; p. 177: Patrick Moore; p. 178: H. Bond (STScI), R. Ciardullo (PSU), WFPC2, HST, NASA; p.179: NASA & ESA; p. 180: California Institute of Technology; p.181: Patrick Moore; p. 183: Subaru Telescope, NAOJ; p. 184: Patrick Moore; p. 185 (top): Patrick Moore; p.185 (bottom) John Bahall, Mike Disney; p. 189: Floyd Clark, California Institute of Technology; pp. 190–1: NASA, ESA, S. Beckwith (STScI) and the HUDF Team.